W0115166

Multiple Sclerosis
Present and Future

Ettore Majorana International Science Series
Series Editor:
Antonino Zichichi
European Physical Society
Geneva, Switzerland

(LIFE SCIENCES)

Recent volumes in the series

Volume 5 **THE AGING BRAIN: Neurological and Mental Disturbances**
Edited by G. Barbagallo-Sangiorgi and A. N. Exton-Smith

Volume 6 **THE LUNG IN ITS ENVIRONMENT**
Edited by G. Bonsignore and G. Cumming

Volume 7 **INVESTIGATION OF BRAIN FUNCTION**
Edited by A. W. Wilkinson

Volume 8 **THE IMMUNOLOGY OF INFANT FEEDING**
Edited by A. W. Wilkinson

Volume 9 **ADVANCES IN NEPHROUROLOGY**
Edited by M. Pavone-Macaluso and P.H. Smith

Volume 10 **CELLULAR BIOLOGY OF THE LUNG**
Edited by G. Cumming and G. Bonsignore

Volume 11 **BIOELECTROCHEMISTRY I: Biological Redox Reactions**
Edited by G. Milazzo and M. Blank

Volume 12 **SELECTED TOPICS IN PREVENTIVE CARDIOLOGY**
Edited by A. Raineri and J. J. Kellermann

Volume 13 **THE BEHAVIOR OF HUMAN INFANTS**
Edited by A. Oliverio and M. Zappella

Volume 14 **DRUGS AND THE LUNG**
Edited by G. Cumming and G. Bonsignore

Volume 15 **AGING AND DRUG THERAPY**
Edited by G. Barbagallo-Sangiorgi and A. N. Exton-Smith

Volume 16 **MULTIPLE SCLEROSIS: Present and Future**
Edited by G. Scarlato and W. B. Matthews

Volume 17 **SMOKING AND THE LUNG**
Edited by G. Cumming and G. Bonsignore

A Continuation Order Plan is available for this series. A continuation order will bring delivery of each new volume immediately upon publication. Volumes are billed only upon actual shipment. For further information please contact the publisher.

Multiple Sclerosis
Present and Future

Edited by

G. Scarlato
Neurological Department
University of Milan
Milan, Italy

and

W. B. Matthews
University of Oxford
Oxford, United Kingdom

Springer Science+Business Media, LLC

Library of Congress Cataloging in Publication Data

International Workshop on Multiple Sclerosis (1983: Erice, Italy)
Multiple sclerosis.

(Ettore Majorana international science series. Life sciences; 16)
"Proceedings of an International Workshop on Multiple Sclerosis, held August 29–September 1, 1983, in Erice, Sicily, Italy"—T.p. verso.
Bibliography: p.
Includes index.
1. Multiple sclerosis—Congresses. 2. Multiple sclerosis—Immunological aspects—Congresses. I. Scarlato, G. II. Matthews, Walter Bryan. III. Title. IV. Series. [DNLM: 1. Multiple Sclerosis—congresses. W1 ET712M v.16 / WL 360 I615m 1983]
RC377.I57 1983 616.8'34 84-18101

ISBN 978-1-4612-9465-8 ISBN 978-1-4613-2403-4 (eBook)
DOI 10.1007/978-1-4613-2403-4

Proceedings of an International Workshop on Multiple Sclerosis, held August 29–September 1, 1983, in Erice, Sicily, Italy

Originally published by Plenum Press, New York 1984
Softcover reprint of the hardcover 1st edition 1984

PREFACE

This book is based on the papers delivered at an International Workshop on Multiple Sclerosis held in Erice, Italy from 29th August to 1st September 1983. The Meeting was organized with great efficiency in the delightful setting of the Centro di Cultura Scientifica Ettore Mahorana, by its Director, Antonino Zichichi, and was directed by the Editors of the Proceedings and by Professor C. Alvisi of the University of Bologna.

The emphasis was deliberately on the contributions of laboratory science to the understanding of multiple sclerosis, its etiology, pathogenesis, diagnosis and treatment. Where so much is unknown, disagreement was expected and indeed welcome, and it is regrettable that it was not possible to publish verbatim the often animated and always interesting discussion.

Differing views on the relevance of experimental allergic encephalomyelitis to multiple sclerosis were expressed by Dr. Wisniewski, Professor Seitelberger and Professor Alvord, who graphically illustrated his changing beliefs over the years. Professor Seitelberger laid much greater emphasis on remyelination in remission in multiple sclerosis than had previously been the accepted view. The disorder of both humoral and cellular immunity in multiple sclerosis was discussed by Dr. Roos and Professor Link. Dr. Gilden described how recombinative techniques might be applied to the detection of a virus in multiple sclerosis. The continued inability to detect any antigen to match the oligoclonal IgG bands in the CSF in multiple sclerosis patients was disappointing and there was an increasing tendency to regard the bands as "nonsense" antibodies. Dr. Latov spoke of the myelin-associated glycoprotein that might be the initial point of attack for the demyelinating process.

Professor Field reiterated challenging views on the laboratory and clinical aspects of multiple sclerosis that have not found wide acceptance.

The difficulties of epidemiological studies were emphasized by Professor Granieri and colleagues, with particular reference to interesting findings of the unexpectedly high incidence in southern Italy.

Diagnostic techniques with particular reference to the pathophysiology of multiple sclerosis and also to its treatment were discussed by Dr. Kolar and Dr. Schauf and the topic of evoked potentials in this context was also reviewed.

Finally, Dr. Gonsette presented a most comprehensive account of treatment of MS based on study of the immunological disorder. It was possible to detect that a number of the authors he cited had tended to confuse evidence of immunosuppression with evidence of clinical benefit, and it was evident that much more precise methods of manipulating the immunological system were required.

G. Scarlato and
W.B. Matthews

CONTENTS

PATHOGENESIS OF PERIVENOUS AND DEMYELINATING ENCEPHALOMYELITIS AND ITS RELEVANCE FOR MULTIPLE SCLEROSIS RESEARCH*

H. M. Wisniewski, H. Lassmann**, G. Schuller-Levis
P. D. Mehta and R. E. Madrid

NYS Institute for Basic Research in Developmental Disabilities
Staten Island, NY 10314

INTRODUCTION

The inflammatory demyelinating plaque in the brain and spinal cord is the hallmark of multiple sclerosis (MS) pathology (Charcot, 1868; Marburg, 1906; Dawson, 1915). The pathology consists of primary demyelination (Marburg, 1906) and chronic perivenous inflammation, mainly located around larger veins and venules (Seitelberger, 1970). It was noted at the beginning of this century that plaques show a high variability of structural changes, i.e., from extensive cellular infiltration to few cells present, various degrees of axonal and oligodendroglia loss, and various degrees of remyelination, gliosis and connective tissue reaction. These observations lead to a large number of different, sometimes even contradictory pathogenetic theories of this disease (Dawson, 1915; Lassmann, 1983).

Experimental allergic encephalomyelitis (EAE), an autoimmune disease of animals induced by challenge with central nervous system (CNS) antigen, had been considered as an experimental model of MS since the earliest descriptions by Rivers et al., (1933). Because the goal of many of these experiments was to determine which component of the brain was responsible for the demyelinating encephalomyelitis, there was a great need for a method which would shorten the time and procedure required to induce the disease. This was achieved when Freund and McDermott (1942) introduced a new technique of immunization by emulsifying antigens in various adjuvants. Using this technique (Kies and Alvord, 1959; Roboz and Henderson, 1959; Laatsch et al., 1962) it was possible to demonstrate that the antigen responsible for EAE was myelin basic protein (MBP). Introduction of Freund's ajuvant (CFA) accelerated the search for finding the antigen responsible for EAE; however, it changed the clinical course and the morphology of the disease. In contrast to the previous chronic, low mortality demyelinating encephalomyelitis, EAE became an acute monophasic disease. The monophasic nonprogressive course of EAE in contrast to progressive and multiphasic course of multiple sclerosis (MS) is one of the main reasons that many investigators do not consider EAE as a model of MS. However, in recent years reproducible small animal models of chronic EAE have become available (Stone and Lerner, 1965; Wisniewski and Keith, 1977; Massanari, 1980; Lublin et al., 1981; Brown et al., 1982). These small animal models cover the full spectrum of MS pathology (Wisniewski et al., 1982; Lassmann, 1983) and allow the study of the pathogenesis of chronic inflammatory demyelinating plaque formation. The present review will focus on new data obtained from these models which may help us understand the mechanisms leading to perivenous and demyelinating encephalomyelitis in humans and animals.

Antigen Recognition in the CNS by Sensitized Lymphocytes

EAE is induced by peripheral sensitization with CNS antigens. Evidence suggests that the disease is initiated by a cellular immune reaction directed against MBP (Waksman and Morrison, 1951; Kies et al., 1960). However, it is not clear how the sensitized T cells, present in the circulation of EAE animals cross the blood brain barrier (BBB) and recognize an antigen(s) which is located in the major dense lines of the myelin sheath. One of the possible mechanisms explaining this phenomenon is that there appears to be a low rate of physiological exchange of lymphocytes between the CNS and the circulation, which allows the entry of the sensitized T cells. The presence of a small number of mononuclear cells (Oemichin, 1983) in the CSF of normal individuals gives some support to this possibility. However, comparing the large number of lymphocytes in the circulation with the small number of hematogenous cells in the normal brain it seems unlikely that sensitized T cells in EAE reach the brain as a result of normal turnover of lymphocytes between the brain and the blood. Another possibility is that sensitized T cells recognize the antigen at the level of the endothelium of cerebral vessels. In a recent study concerning experimental subacute sclerosing panencephalitis (SSPE) we have shown that virus antigen is present in the cytoplasm and on the luminal surface of endothelial cells of cerebral vessels in infected animals (Wisniewski et al., 1983a,b). This viral antigen may be the target of the initial inflammatory reaction in this disease. In EAE, recognition of the antigen at the level of the endothelium of cerebral vessels is only possible if the following requirements are fulfilled: 1) the presence of a soluble pool of MBP in the extracellular space of the brain; 2) the transport of MBP through the endothelial barrier of cerebral vessels; and 3) the presentation of MBP together with histocompatibility antigens on the luminal surface of cerebral vessels. Recently we have shown that MBP injected into the CSF of normal rats is transported through the endothelial barrier of meningeal vessels, and is then bound to the luminal surface of endothelial cells of meningeal veins (Vass et al., 1983). Furthermore the presence of immune response associated (Ia) antigen on the surface of cerebral endothelial cells has been documented in normal as well as EAE animals (Sobel, 1983). These results indicate that MBP, when liberated in the CNS tissue may be transported through the BBB and presented on the luminal surface of cerebral vessel endothelium. The question remains unsettled, whether the release of MBP in normal animals is sufficient to allow recognition of this antigen on the endothelial surface. The presence of an MBP serum factor in normal and EAE animals has been described by Paterson and Day (1979), which may originate from MBP turnover in the CNS tissue. Furthermore it has to be considered, that minor traumatic events in the CNS tissue may liberate small amounts of MBP. This could explain the high incidence of lesions in EAE and MS in areas of the brain and cord with a high probability of minor mechanic strain, like tearing and shearing forces (Oppenheimer, 1978; Lassmann, 1983).

Factors Governing the Development of Acute and Chronic Disease

Clinico-pathological studies using various doses of antigens suggest that the induction of a chronic type of EAE depends upon a delicate balance between the antigen dose and the host response (Lassmann and Wisniewski, 1979; Wisniewski and Madrid, 1983; Lassmann, 1983). For example, sensitization of young guinea pigs with 25 mg spinal cord tissue and 1 mg *Mycobacterium tuberculosis* leads to severe and fatal acute EAE. Doubling the encephalitogenic dose in the same animals reduces the acute mortality and leads to a high incidence of chronic EAE. Further increase of the dose leads to further reduction of mortality in acute EAE, but also decreases the severity of chronic EAE. Challenging adult animals with a similar dose of encephalitogen as necessary for the induction of chronic relapsing EAE in juvenile animals results in high mortality during acute EAE (Lassmann and Wisniewski, 1979). However, chronic progressive EAE may be induced in adult guinea pigs with a very high dose of encephalitogen (Wisniewski and Madrid, 1983). Thus the induction of chronic EAE may be due to suppression of acute EAE with high doses of antigen, which however are not high enough to induce full high zone tolerance.

Continuous Antigenic Stimulation - A Requirement for Chronic Disease Activity

The factors responsible for the establishment of chronic progressive or recurrent disease in MS pathogenesis are unresolved. Observations based on both humans and animals suggest the necessity of a peristent peripheral antigenic stimulation for development of chronic inflammation and demyelination (Wisniewski et al., 1982). These observations were confirmed by Tabira et al., (1983) who, after removal of the inoculum by leg amputation, observed arrest of chronic EAE. Exactly how a peripheral deposit of neuroantigen acts on the maintenance of the disease process in chronic EAE is at present unknown. It is possible that a balance between release and local degradation of the neuroantigen is established through the action of local factors at the site of injection. If at one point in time a mitogenic dose of neuroantigen reaches the draining lymph nodes, then proliferation of specific neuroantigen-sensitized cells would occur and a clinical relapse would follow. A recent report describing the induction of relapsing EAE in mice after five-six months of adoptive transfer of syngeneic lymph node cells from MBP-sensitized animals would appear to contradict the prerequisite of continuous or discontinuous antigenic stimulation for the induction of chronic relapsing EAE (Mokhtarian et al., 1983). The mechanism of induction of chronic disease by adoptive transfer is as yet unclear.

In this regard, interesting observations have been made in humans during the study of neurological complications after rabies vaccination (Uchimura and Shiraki, 1957), and as a result of therapeutic trials with brain tissue (Seitelberger et al., 1967). In both situations, an inflammatory demyelinating disease of the CNS, closely resembling acute MS, was induced during the course of injections. However, the disease dissipated weeks to months after the last injection of brain tissue (Haruhara, 1958). There is also no evidence that such pathological processes of the brain as inflammation, tumor, trauma or stroke release autoantigens sufficient to develop perivenous or demyelinating encephalomyelitis. These observations suggest that in humans autosensitization leading to demyelinating disease does not occur. It is possible that in humans, like in experimental animals, discontinuous or continuous peripheral antigenic stimulation may be necessary for maintenance of a chronic inflammatory demyelinating disease.

The Humoral Immune Response in EAE and its Relation to Disease Activity

One of the important findings in MS is the increased level of IgG in CSF (Tourtellotte, 1972). Comparison of IgG to albumin ratios in CSF and blood (CSF IgG index) indicates that IgG is synthesized within the CNS in MS (Ganrot and Laurell, 1974; Olsson and Pettersson, 1976). A number of investigators (Mehta et al., 1981; Karcher et al., 1982) have shown that guinea pigs with acute and relapsing EAE have increased concentrations of IgG in their CSF or in neutral pH extracts of brain tissue. Employing the CSF IgG index, some investigators (Glynn et al., 1982; Suckling et al., 1983) were unable to determine whether IgG synthesis took place within the CNS of relapsing EAE animals. BBB damage which accompanies active demyelinating lesions was probably a contributing factor in the difficulty in determining whether IgG was synthesized within the CNS or entered the brain as a result of BBB changes. In contrast, recent studies (Karcher et al., 1983; Olsson et al., 1983) showed that animals which were sacrificed after 125 days, and demonstrated mild clinical signs and symptoms as well as few active lesions had evidence of intrathecal synthesis of IgG. Immunofluorescence and electron microscopic studies (Grundke-Iqbal et al., 1980; Madrid et al., 1981) have shown the presence of plasma cells within the parenchyma in relapsing EAE animals supporting the notion that part of the IgG may be synthesized within the CNS.

A number of investigators using electrophoresis or isoelectric focusing demonstrated the presence of oligoclonal IgG bands in CSF from 85 to 90% in patients with MS (Johnson et al., 1977; Laurenzi et al., 1980; Lowenthal, 1979; Mattson et al., 1981; Mehta et al., 1981). Evidence of bands in serum, CSF and CNS extracts in relapsing EAE was also found (Mehta et al., 1981; Olsson et al., 1982; Karcher et al., 1983; Glynn et al., 1982). Recent studies have shown that in chronic relapsing EAE animals, the oligoclonal IgG band patterns were identical in CSF and serum (Mehta et al., 1983). These results suggest that the plasma cells present within and outside the CNS are of the same type. In contrast, about 50% of MS sera did not show oligoclonal bands and those which showed bands had patterns either partially similar or completely different from those seen in matching CSF (Mehta et al., 1982). The reason for the discrepancy at present is not clear, however, it could be they are a response to (a) antigen(s) present within the CNS or (b) they could be masked by polyclonal IgG. Thus, it seems that in MS, the CSF oligoclonal IgG represents a response to antigen(s) present within the CNS, whereas in serum the bands are probably masked by polyclonal IgG or presence of immune complexes. The chronic EAE animals have peripheral deposits of CNS antigens. Therefore, the pattern of oligoclonal bands in the brain extracts, CSF and serum is similar. In MS the location of the antigens driving the disease are not known.

The specificity of the major oligoclonal IgG in MS is not known. However, in most MS patients, evidence of the intrathecal IgG synthesis to a number of viruses (Vartdal et al., 1980; Salmi et al., 1983) as well as the presence of anti-brain antibodies in their CSF (Ryberg, 1978) has been reported. It is possible that the major oligoclonal IgG bands are not directed against any specific set of antigens but represent a general activation of B-cell clones during attacks of MS (Paterson and Whitacre, 1981). A similar situation may exist in the sera of relapsing EAE animals, since a majority of oligoclonal bands in sera of chronic relapsing EAE animals were not directed against CNS antigens (Glynn et al., 1982). Although a number of studies showed antibody to MBP in CSF and sera, and antibody to glycolipid and myelin in sera (Schwerer et al., 1983), it is not clear if the activity is associated with oligoclonal or polyclonal IgG.

In summary, increased IgG levels and oligoclonal bands are present in both chronic EAE and MS. Moreover, they do not seem to change during remissions and exacerbations in both MS and relapsing EAE. In chronic-EAE the oligoclonal bands (in spite of the fact that CNS antigens were used for sensitization) are not directed solely to CNS antigens. Without the knowledge of the sensitizing antigen in chronic EAE, analysis of oligoclonal brand and the levels of IgG, in spite of existing deposits of these antigens, did not tell us that the disease was initiated and driven by antigens of CNS origin. If we cannot determine the antigens to which the animal is sensitized using a powerful adjuvant, strong encephalitogenic antigens and sensitive immunologic techniques, the possibility of determining the driving antigens for the initial attack and relapse in MS using the same techniques, in our opinion is hopeless. Data from animal studies have shown that of many CSF oligoclonal bands, only a few can be absorbed with CNS antigens (Glynn et al., 1982). It is not clear whether these non-CNS antibodies play any role in the pathogenesis of tissue damage. THeoretically all anti-brain antibodies (see mechanism of tissue damage) and antibodies directed toward foreign antigens residing in the CNS can contribute to the process of demyelination, oligodendroglia and axonal loss. However, available methodologies have not allowed the determination of antibodies which may be important during a new attack of the disease.

The Cellular Immune Response in Chronic EAE and its Relation to Disease Activity

Cell-mediated immune (CMI) mechanisms are known to be largely responsible for allograft rejection, microbial immunity of the tuberculin type as well as the recognition of a variety of simple chemical and possibly tumor antigens. The role of CMI in human autoimmune disease is less clearly established, but sufficient evidence suggests an important component and even central role of CMI in experimental autoimmune thyroiditis, orchitis, myasthenia gravis, and EAE (Neuwelt and Clark, 1978).

There are numerous studies of CMI in animals with acute EAE. Relatively few studies have investigated CMI in a chronic relapsing model. Traugott et al., (1978) have reported a decrease in circulating early T cells (using rosetting techniques) associated with both the onset of disease and with subsequent relapses. It is not clear, however, whether MBP-reactive T cells are responsible for relapses in chronic EAE.

Our results indicate that a peripheral T cell response to MBP correlates with the onset of the first attack of chronic EAE. This time course parallels that of skin-reactivity to MBP (Shaw, 1965) and that of acute circulating T cells (Traugott, 1978) in acute EAE. Neither the T cell response, as measured by lymphocyte transformation to MBP, nor serum MBP-antibody level seem to correlate with relapses and remissions in chronic EAE (Schwerer et al., 1983). Thus, the involvement of MBP in the precipitation of relapses remains unclear. On the other hand, as discussed above, CNS-antigens other than MBP might be responsible for relapses and demyelination in chronic EAE. CMI studies do not conclusively show that immune reactions against MBP or other CNS-antigens play a role in the pathogenesis of MS.

Most studies of CMI in EAE have used lymphocytes obtained from extraneural sites. By studying the properties and functions of inflammatory cells within the CNS lesion, one would expect to better understand the relationship of lymphoid cells to disease activity. One such study by Lublin and Maurer (1981) was aimed at showing the differences in the immune response of lymphocytes from the CNS of guinea pigs with acute EAE compared to peripheral blood lymphocytes from these animals. They reported no proliferative response when the CNS lymphocytes were cultured with MBP. THere are several possible explanations for this lack of response. There may be factors (suppressor macrophages or lymphocytes) which render the isolated cells unresponsive to MBP (see discussion below). Also myelin fragments from the brain tissue may saturate the receptors for MBP. It may also be possible that too few MBP responsive T cells were isolated by this technique.

Our laboratory has also studied (Schuller-Levis et al., 1983) inflammatory cells isolated from the CNS of guinea pigs with chronic EAE. Similar to the results of Lublin and Maurer (1981), preliminary findings show a minimal response of these lymphocytes to MBP, again showing no evidence for the relationship of disease activity and CMI under these conditions.

With the standard proliferation assay, up to now it has not been possible to detect a significant proliferative response to MBP from both peripheral and brain lymphocytes in chronic EAE models. The studies of Ben-Nun and Lando (1983) in an acute EAE mouse model have shown that the removal of the adherent cell fraction from the immunized cell population and its replacement with fresh naive accessory cells could reveal a significant proliferative response to MBP. THese findings indicate that MBP responsive T cells may be present during the course of chronic EAE, but are in an unresponsiveness state under the conditions employed. The reasons for the unresponsiveness to MBP requires further studies and may be important in understanding the course of chronic EAE. MBP antigen excess with subsequent high dose tolerance and activation of T cell suppression also has to be seriously considered. T cell activation is known to require syngeneic Ia products plus antigen (Schevach and Rosenthal, 1973; Benacerraf and McDevitt, 1972). The T cell recognizes the Ia product and the MBP epitope. Studies with synthetic polypeptide antigens (Kapp et al., 1975; Benaceraff and McDevitt, 1975) have shown that polypeptide antigens, free from Ia containing cells, will activate T suppressor mechanisms. If the same principle holds for MBP or other polypeptides possibly involved in chronic EAE, similar mechanisms may be implicated. The finding of fresh Ia bearing cells restoring unresponsiveness suggests that an equilibrium exists between the accessory cell Ia saturated or unsaturated with MBP and the T cell with its Ia and MBP receptors. These data suggest that the state of responsiveness or unresponsiveness of the T cell may be important in the pathogenesis of EAE. Future studies, rather than concentrating on the standard assessment of CMI during the course of EAE should focus attention on the reasons behind the unresponsive state of the T cell.

The abnormalities of cell mediated immunity in MS have been well reviewed by Waksman (1981) and McFarlin and McFarland (1982). Total T and B cells vary little with the long term course of the disease. However, at the time of acute exacerbation a slight fall in T cells has been shown by the use of rosetting techniques and monoclonal antibodies (Lisak, 1975; Reinherz, 1980). Functional changes have also been reported in peripheral T cells with regard to responsiveness to mitogens. Also, in one study natural killer (NK) cells and NK precursors induced by interferon were deficient in a third or more of the MS patients (Bloom, 1980). Peripheral blood leukocytes have been shown to have a marked increase in antibody-dependent cellular cytotoxicity (Merrill 1982). There are also many other changes in CMI reported in MS. However, most of these subtle changes are not clearly related to clinical disease activity, and may also occur in other neurological diseases and normal individuals.

As indicated above, T and B cells and macrophages do play an important role in the pathogenesis of MS, although changes in these cells do not always correlate with disease activity. Problems emerge in evaluating CMI data due to the high variability of the disease from patient-to-patient, lack of longitudinal studies, different methods of evaluation, problems of quantifying _in vitro_ assays, and variation of therapy. The changes on CMI produced by steroids alone cannot be ignored. Most investigations of CMI in MS have been done using peripheral lymphoid cells, because of the difficulty in obtaining CNS lymphocytes. It is important to emphasize that CMI changes in the peripheral lymphoid system are subtle and that studies of lymphocyte abnormalities in the CNS could be more promising. However, thus far, alterations in CMI have not clearly identified the primary event (autoimmune, virus) which leads to the development of MS nor have these changes correlated closely with disease activity.

Inflammatory and Demyelinating Antigens

Sensitization studies with various fractions of CNS tissue indicate that independent antigens may be responsible for inflammation and demyelination (Figure 1). Sensitization with purified MBP may lead to acute or chronic encephalomyelitis. This disease however is predominantly inflammatory with perivenous distribution and little demyelinating activity (Pannitch and Ciccone, 1981, Madrid et al., 1982). Recent studies show that there are other CNS antigens like lipophilin and proteolipid apoprotein (Hashim et al., 1980; Madrid et al., 1982; Cambi et al., 1983) which can also induce perivenous encephalomyelitis. A combined sensitization with MBP and antigens like galactocerebroside or sialogalactosylceramide (GM4) leads to the formation of large confluent demyelinating plaques (Madrid et al., 1983). The concept of two classes of antigen needed for induction of demyelinating plaques is consistent with our previous observations (Madrid et al., 1981; Wisniewski et al., 1982) that polyantigenic stimulation was most effective in the development of extensive demyelination.

Mechanisms of Tissue Damage

Antibody dependent cell mediated cytotoxicity (ADCC) and bystander demyelination have been proposed (Wisniewski and Bloom 1975; Wisniewski, 1977; Brosnan et al., 1977) as a leading mechanism responsible for tissue damage and plaque formation in inflammatory demyelinating diseases. These as well as other observations on the role of macrophage activation products, complement and direct cellular attack were recently reviewed (Wisniewski and Lassmann, 1983). In this section some of the new information not included in these reviews is discussed.

Recent data by Trotter and Smith (1983) have demonstrated that peritoneal macrophages secrete phospholipases and that macrophage homogenates degrade myelin *in vitro*. Opsonization of myelin resulted in a 50% to 300% increase in the uptake of myelin by monolayers of macrophages.

There is also recent evidence for a direct cellular destruction of oligodendrocytes *in vitro* by T lymphocytes sensitized against galactocerebrosides (Niediek and Lehman, 1982). However, there is overall little evidence for direct cellular attack in the absence of humoral factors.

Several recent experiments have focused on the role of serum factors and complement in demyelination. In sera from rabbits with white matter-induced-EAE, Grundke-Iqbal et al., (1981) have demonstrated that in addition to IgG complement-dependent factors non-IgG proteins are responsible for *in vitro* demyelination. In addition,

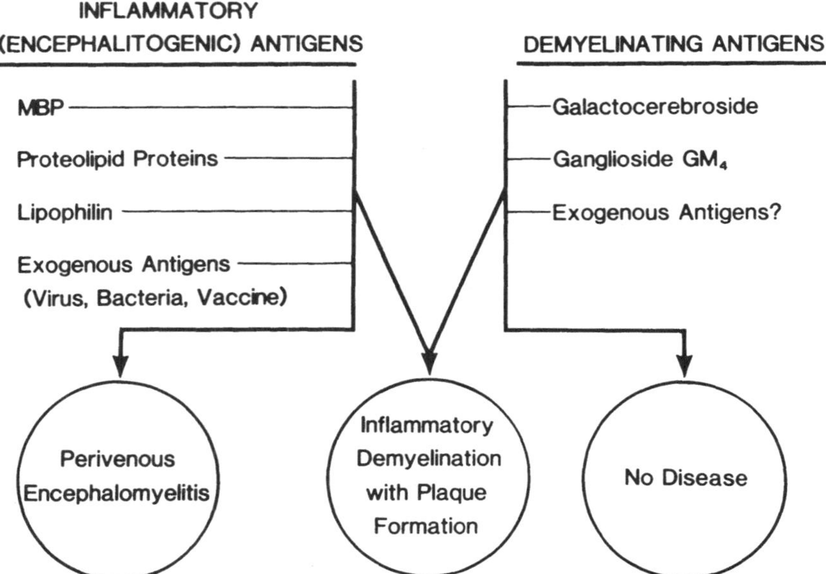

Fig. 1. Inflammatory and demyelinating antigens and their role in pathogenesis of perivenous and demyelinating encephalomyelitis.

data indicate that, unlike acute EAE, in MS most of the serum demyelinating activity is due to non-immunoglobulin factors, which are possibly complement dependent, and that serum immunoglobulins play a minor role in demyelination in vitro (Bornstein and Grundke-Iqbal, 1982).

The exact interaction of cells and serum factors at the site of inflammation which lead to demyelination remains unknown; indeed, understanding this process may eventually lead to modification of tissue damage by pharmacologic agents.

Multiple Sclerosis as an Autoimmune Disease

Usually in the immune system some lymphocytes are present which are capable of recognizing autoantigen (Weckerle et al., 1974). Thus in normal individuals a low level of autoreactivity against CNS antigens like MBP, cerebroside and ganglioside is frequently present (Yung et al., 1973; Diessner et al., 1974; Dore-Duffy et al., 1980; Frick, 1982). The manifestation of an autoimmune disorder appears to be prohibited by immunoregulatory mechanisms. Recent immunological studies in MS have shown an apparent defect in immune regulation (Waksman, 1981; Weiner and Hauser, 1982), most importantly a decrease of suppressor cells in the blood and CSF of patients during activity of the disease (Antel et al., 1979). It is interesting that in a recent study all MS cases with active disease demonstrated a high cytotoxicity of T-lymphocytes directed against MBP, encephalitogenic peptide, cerebroside, as well as gangliosides (Frick, 1982). However, earlier studies, using different test systems, did not show good evidence for autoimmunity in MS (Weiner and Hauser, 1982). Furthermore, treatment trials with MBP in MS thus far have not revealed clear-cut positive results (Salk et al., 1980). The pathogenetic role of MBP and other autoantigens in MS is therefore still controversial.

The Role of Hetero-Antigens in the Etiology of MS

Although, as discussed before, autoimmunity appears to play an important role in the pathogenesis of inflammatory demyelinating lesions, several aspects of MS and related inflammatory demyelinating diseases are difficult to explain on the basis of autoimmunity alone. One important aspect deals with the induction and maintenance of the disease process. As discussed above in chronic EAE the disease is arrested within a few weeks after loss or removal of the sensitizing antigen from the inoculation site (Wisniewski et al., 1982; Tabira et al., 1982). A similar continuous peripheral antigenic stimulation may also be required for MS. In humans, a disease closely resembling acute MS (Marburg, 1906) is induced by active sensitization of patients with brain tissue (Uchimura and Shiraki, 1957; Seitelberger et al., 1958). However, in all described cases the disease process ceased a few months after the last injection of antigen. Furthermore, there is at present no evidence available that damage to CNS tissue with liberation of CNS antigens may by itself induce a self perpetuating autoimmune disease directed against myelin or oligodendroglia.

Indirect epidemiological evidence suggests that MS may be induced by an exogenous agent. At this point it should be noted that the many attempts to demonstrate the presence of virus in the brain of MS patients, using tissue culture and/or morphological techniques (EM, immunohistochemistry) have failed to find virus. However, absence of evidence is not evidence of absence, and there are nucleic acid hybridization studies which suggest that virus material may be present in the CNS of MS patients (Weiner et al., 1973; Simon, 1982).

There are several ways in which exogenous agents may be involved in the etiology and pathogenesis of MS. A virus may, by infecting lymphoid cells, interfere with immune regulation. This may allow the action of autoreactive clones of lymphoid cells directed against CNS antigens. It is difficult to understand however how selective autosensitization against CNS tissue can occur. One of the theoretical possibilities is that components of self, especially Ia product, are involved in T cell recognition and that different Ia plus different environmental antigens may result in organ specific attack in individuals of selected histocompatibility. Viruses may also induce an autoimmune reaction against CNS antigens by cross-reactivity or by transient infection of oligodendrocytes. However, currently there is no evidence for cross-reactivity of viral and brain antigens and no evidence that autoimmune disease can be established as a result of oligodendroglia infection. Finally, it is possible that the genomic infection of the CNS by the virus may result in an inflammatory demyelinating disease morphologically indistinguishable from that resulting from autosensitization. This possibility is under current study in several laboratories.

SUMMARY

In a summary statement of the Report and Recommendations of the National Commission on Multiple Sclerosis published in 1974 it was said "One obstacle in the path of investigators seeking the cause of and cures and treatments for multiple sclerosis is the fact that the disease occurs, so far as is yet known, only in man", and "what MS researchers need for the ultimate benefit of patients is a susceptible animal - in laboratory language an animal model.... So far, either the appropriate animal has not been selected or a correct method of inducing the disease has not been found." At present, the closest approach to an animal model of MS is EAE to which many species of laboratory animals are susceptible. Approximately $1,000,000 a year is being spent on EAE research, primarily in the hope that it will provide clues to multiple sclerosis. However, there is controversy among investigators as to the value of this model. Advocates of the EAE model point to its resemblence to MS in one important respect: it results in loss of myelin in the CNS - the amount varying with the species of animal used. EAE can be induced at will in a susceptible strain of animals by the injection of any one of a number of antigenic proteins; it can be almost as readily switched off by an injection of the same material that induced the disease. Critics of EAE point to the important way in which it differs from the human disease: when it is allowed to run its course, EAE is not marked by the unpredictable cycles of exacerbation and remission so characteristic of MS. The National Advisory Commission took a "judgment reserved" position in regard to the value of this experimental model of MS.

A Satellite Conference on EAE was held July 16-19, 1983 in Seattle, Washington, and the question asked by Dr. E. C. Alvord was, "How good a model of MS is EAE today?" Dr. E. C. Alvord, one of the most productive contributors in the studies of EAE, during the welcoming remarks showed his graph illustrating, since 1944, the ups and downs regarding suitability of EAE as a model of MS. He stated, "I am 80% convinced that EAE is the best model of MS. Which way are you going to move me?"

In our opinion recent data does not call for an up or down shift of the graph but rather for a change of the graph. From the studies of the autoimmune chronic relapsing and/or chronic progressive EAE, it appears that the clinical course of the disease, immunology, electrophysiology, and pathology is similar to that seen in MS (Table 1). Therefore, for the study of the pathogenesis of MS the pure autoimmune chronic relapsing and progressive model of EAE in guinea pigs and rats should score almost 100% on Dr. Alvord's graph. However, there is growing evidence that the initiating antigen in MS is not MBP or any other antigen(s) of CNS origin. For example, none of the lesions releasing a lot of autoantigens such as trauma, stroke and tumors of the brain, lead to MS-like pathology. Therefore, from an etiological point of view MS and chronic EAE are diseases caused by very different antigens.

What are some of the lessons that have been elucidated from the autoimmune acute and chronic relapsing and/or progressive EAE models:

1. CNS tissue contains both inflammatory and demyelinating antigens.
2. Histocompatibility antigens and various levels of tolerance govern the susceptibility and development of acute and/or chronic relapsing or progressive autoimmune EAE.
3. Continuous or discontinuous antigenic stimulation is needed for the development of chronic relapsing or progressive EAE.
4. There is no evidence that naturally occurring pathological processes in the brain (inflammation, tumor, mechanical trauma, stroke) lead to the release of autoantigens sufficient to initiate or to continue an autoimmune perivenous or demyelinating encephalomyelitis.
5. Presence of anti-brain antibodies, (whatever the reason for their appearance) in the presence of macrophages may lead to demyelination, oligodendroglia and axonal loss.
6. In chronic EAE current humoral and cell mediated immunologic diagnostic techniques have not allowed us to determine the antigens to which the animal was sensitized. In practical terms it means that in chronic EAE, immunological methods, which in infectious diseases can help to identify the infectious agent and determine the activity of the disease, did not clearly identify the antigen(s) responsible for initiation of the disease. This conclusion is disturbing because current immunological techniques are unable during the chronic stage of autoimmune EAE to determine the antigens which were used for sensitization and of which deposits still are present at the site of injection. If this is true in experimental situations it is probably also true in MS. Therefore with immunological technologies discussed above, it is unlikely that either the antigens responsible for initiating MS nor the antigens responsible for relapses in MS will be determined.

Table 1. Similarities of the Chronic Relapsing EAE Model to MS.

Clinically	Chronic relapsing progressive course.
Pathology	Large (macroscopically visible) demyelinated plaques with comparable topographical distribution. Inflammation (mononuclear cells). Primary demyelination. Oligodendrocyte loss. Reactive gliosis. Simultaneous presence of plaques of different age. Plaque growth by either confluence of perivenous lesions or radial growth in preexisting plaques.
Immunology	Involvement of both cellular and humoral immune systems in the pathogenesis of the disease.
Electrophysiology	Dominance of delayed conduction velocity in VEP. Some reduction in amplitudes.
Other	Critical age of dependency. Genetic susceptibility.

Should the Inducing Antigen(s) (Etiology) of MS be Identified Today; How Effectively could the DIsease be Treated?

1. If MS is a pure autoimmune disease, no specific (i.e., desensitization) therapy is effective in other known autoimmune diseases (e.g., thyroiditis). A broad spectrum of immunosuppressive and anti-inflammatory therapies was currently in use.

2. If MS is caused by exogenous (hetero) antigen(s), e.g., persistent or genomic viral infection, today there is no specific antiviral therapy. Experience with SSPE using antimeasles antibodies or vaccine during the advanced stage of disease does not appear to be effective.

3. If MS is a result of a combination of the autoimmune and exo-antigen(s) hypothesis, again, there is no effective therapy.

Historically, knowledge of the cause and/or pathogenesis of a given disease leads to cure or a rational therapy. Experience with slow infections caused by well known conventional viruses like SSPE indicate that so far a therapeutic strategy to treat the patient once the signs and symptoms of the disease occur. However, if vaccination programs are successful then chances of eradicating the disease are good. Even a superficial review of the treatment regimes of autoimmune diseases indicate that we did not succeed in development of any specific therapy. Why did it happen? The reason is the lack of basic information about the immune system mechanism rejecting or allowing infectious agents or any other antigen(s) to interact with host cells (histocompatibility antigens, mechanism of low and high tolerance, role of interferons and interferon like molecules etc). Our poor understanding of the mechanism of viral genomic and persistent infections is another factor which does not allow the development of successful antiviral therapy.

Recognizing that not enough is known about the basic processes which lead to the development of MS, what can be done to improve the chances of effectively treating MS patients? On the basis of the data discussed in this review, we think it is possible to identify areas of the most promising research in MS rather than leave it to a chance phenomenon - which appears to be done by some of the MS research funding agencies. While discussion on specific priorities of research in MS is beyond the scope of this review, we do elaborate these issues in a paper in preparation: "Current Trends in MS Research - Areas of Priorities."

* To Mrs. Alma Sagan, the best fighter of the MS malady and all those heros who fight back the disease, this paper is dedicated.

** Dr. Hans Lassmann is a visiting scientist from the Neurological Institute, University of Vienna, Vienna, Austria.

Acknowledgement

Part of the work described in this manuscript has been supported by USPHS grant NS 14406.

REFERENCES

Alvord, E. C., 1983, Historical perspective and challenge: How good a model of Multiple Sclerosis is experimental allergic encephalomyelitis today? Abstract from the Satellite Meeting of the International Society of Neurochemistry, Seattle, Washington.

Antel, J. P., Arnasan, B. G. W., and Medof, M.E., 1979, Suppressor cell function in multiple sclerosis: correlation with clinical disease activity, Ann.Neurol., 5:338.

Benacerraf, B., and McDevitt, H., 1972, Histompatibility-linked immune response genes, Science 175:273.

Ben-Nun, A., and Lando, Z., 1983, Detection of autoimmune cells proliferating to myelin basic protein and selection of T cell lines that mediate experimental autoimmune encephalomyelitis (EAE) in mice, J.Immunol., 130:1205.

Bloom, B., 1980, Immunological changes in multiple sclerosis, Nature (London) 287:275.

Bornstein, M. B., and Grundke-Iqbal, I., 1982, Circulating meylinotoxic and neuroelectric blocking factors in demyelinative disorders, in: "Clinics in Immunology and Allergy," W. B. Waksman, eds., W. B. Saunders Co., Ltd., Philadelphia, Vol. 2, p.297.

Brown, A., McFarlin, D. E., and Raine, C. S., 1982, Chronologic neuropathology of relapsing experimental allergic encephalomyelitis in the mouse, Lab.Invest, 46:171.

Cambi, F., Lees, M. B., Williams, R. M., and Macklin, W. B., 1983, Chronic EAE produced by bovine protolipid apoprotein. Immunologic studies in rabbits, Ann.Neurol., 13:303.

Charcot, J. M., 1868, Histologie de la sclerose en plaque, Gazette hopital (Paris) 41:554.

Dawson, J. W., 1915, The histology of disseminated sclerosis, Trans.Royal.Soc.Edinb., 50(3):517.

Diessner, H., Neumann, V., and Schmitdt, R. M., 1974, Zur Einwirkung von J-125 markiertem encephalitogenen Progein auf periphere Blutylmphozyten, Schweiz.Arch.Neurol., 114:57.

Dore-Duffy, P., Goertz, V., and Rothman, B. L., 1980, Lymphocyte adherence to myelinated tissue in multiple sclerosis, J.Clin.Invest., 66:843.

Freund, J., and McDermott, K., 1942, Sensitization to horse serum by means of adjuvants, Proc.Soc.Exp.Biol. & Med., 49:548.

Frick, E., 1982, Cell mediated toxicity by peripheral blood lympocytes against basic protein of myelin, encephalitogenic peptide, cerebrosides and gangliosides in multiple sclerosis, J.Neurol.Sci., 57:55.

Ganrot, K., and Laurell, C. B., 1974, Measurement of IgG and albumin content of cerebrospinal fluid, and its interpretation, Clin. Chem., 2:571.

Glynn, P., Weedon, D., Edwards, J., Suckling, A. J., and Cuzner, M. L., 1982, Humoral immunity in chronic relapsing experimental autoimmune encephalomyelitis, J.Neurol.Sci., 57:369.

Grundke-Iqbal, I., and Bornstein, M. B., 1980, Multiple sclerosis: Serum gamma globulin and demyelination in organ culture, Neurol., 30:749.

Grundke-Iqbal, I., Lassmann, H., and Wisniewski, H. M., 1980, Immunohistochemical studies in chronic relapsing experimental allergic encephalomyelitis, Arch.Neurol., 37:651.

Grundke-Iqbal, I., Raine, C. S., Johnson, A. B., Brosnan, C. F., and Bornstein, M. B., 1981, Experimental allergic encephalomyelitis: Characterization of serum factors causing demyelination and swelling of myelin, J.Neurol.Sci., 50:63.

Haruhara, C., 1958, Clinical observations on encephalitis and its residual symptoms caused by anti-rabies treatment, Seishin Shinkeigaku Zasshi, Psychiatria et Neurologia Japonica, 58:355.

Hashim, G. A., Wood, D. D., and Moscarello, M. A., 1980, Myelin lipophilin-induced experimental allergic encephalomyelitis in guinea pigs, in: "Neurochemistry and Clinical Neurology," L.

Battistin, G. A. Hashim and A. Lajtha, eds., A. R. Liss, New York, p.21.

Johnson, K. P., Arrigo, S. C., Nelson, B. J., and Ginsberg, A., 1977, Agarose electrophoresis of cerebrospinal fluid in multiple sclerosis. A simplified method for demonstrating cerebrospinal fluid oligoclonal immunoglobulin bands, Neurology., 27:273.

Kapp, J., Pierce, C., and Benacerraf, B., 1975, Genetic control of immune responses in vivo. IV. Experimental conditions for the development of helper T cell activity specific for the terpolymer L-glutamic acid^{60}-L-alanine^{30}-L-tyrosine^{10}(GAT) in nonresponder mice, J.Exp.Med., 142:50.

Karcher, D., Lassmann, H., Lowenthal, A., Kitz K., and Wisniewksi, H. M., 1982, Antibodies - restricted heterogeneity in serum and cerebrospinal fluid of chronic relapsing experimental allergic encephalomyelitis, J.Neuroimmunol., 2:93.

Kies, M. W., and Alvord, E. C., Jr., 1959, "Allergic Encephalomyelitis," Charles C. Thomas, Springfield, Illinois.

Kies, M. W., Murphey, J. B., and Alvord, E. C., 1960, Fractionation of guinea pig brain proteins with encephilitogenic activity, Fed.Proc., 19:207.

Lassmann, H., 1983, A Comparative Neuropathology of Chronic Experimental Allergic Encephalomyelitis and Multiple Sclerosis, Neurology Series, Vol. 25, Springer, Berlin-Heidelberg-New York.

Lassmann, H., Karcher, D., Kitz, K., Lowenthal, A., and Verslegers, W., 1983, Humoral immune response in guinea pigs at different stages of EAE. Abstract from the Satellite Meeting of the International Society of Neurochem., Seattle, Washington.

Lassmann, H., Stemberger, H., Kitz, K., and Wisniewski, H. M., 1983, In vivo demyelinating activity of sera from animals with chronic experimental allergic encephalomyelitis: Antibody nature of the demyelinating factor and the role of complement, J.Neurol.Sci., 59:123.

Lassmann, H., and Wisniewski, H. M., 1979, Chronic relapsing experimental allergic encephalomyelitis. Clinicopathological comparison with multiple sclerosis, Arch.Neurol., 36:490.

Lassmann, H., and Wisniewski, H. M., 1979, Chronic relapsing experimental allergic encephalomyelitis: Effect of age at the time of sensitization on clinical course and pathology, Acta Neuropathol., 47:111.

Laatsch, R. H., Kies, M. W., Gordon, S., and Alvord, E. C., Jr., 1962, The encephalomyelitic activity of myelin isolated by ultracentrifugation, J.Exp.Med., 115:777.

Laurenzi, M. A., Mavra, M., Kam-Hansen, S., and Link, H., 1980, Oligoclonal IgG and free light chains in multiple sclerosis demonstrated by thin-layer polyacrylamide gel isoelectric focusing and immunofixation, Ann.Neurol., 8:241.

Lisak, R., Levinson, A., Zweiman, B., and Abdou, N., 1975, T and B lymphocytes in multiple sclerosis, Clin.Exp.Immunol., 22:30.

Lowenthal, A., 1979, Restricted heterogeneity of the IgG in neurology, in: "Humoral Immunity in Neurological Diseases," D. Karcher, A. Lowenthal, A. D. Strosberg, eds., Plenum Press, New York, pp. 281-288.

Lublin, F. D., and Maurer, P. H., 1980, Separation and properties of lymphocytes obtained from the central nervous system, J.Clin. Lab.Immunol., 4:195.

Madrid, R. E., Wisniewski, H. M., Hashim, G. A., Moscarello, M. A., and Wood, D. D., 1982, Lipophilin-induced experimental allergic encephalomyelitis in guinea pigs, J.Neurosci.Res., 7:203.

Madrid, R. E., Wisniewski, H. M., Iqbal, K., Pullarkat, R. K., and Lassmann, H., 1981, Relapsing experimental allergic encephalomyelitis induced with isolated myelin lipids, J.Neurol.Sci., 50:399.

Madrid, R. E., Yu, R. K., Deshmukh, D., and Wisniewski, H. M., 1983, Chronic relapsing and progressive EAE induced by supraoptimal immunization with MBP and galactocerebroside, J.Neuropath Exp.Neurol., 42:358.

Marburg, O., 1906, Die sogenannte "akute multiple sclerose." Jahrb.f.Psych., 27:211.

Massanari, R. M., 1980, A latent-relapsing neuroautoimmune disease in Syrian hamsters, Clin.Immunol.Immunopath., 16:211.

Mattson, D. H., Roos, R. P., and Arnason, B. G. W., 1981, Comparison of agar gel electrophoresis and isoelectric focusing in multiple sclerosis and subacute sclerosing panencephalitis, Ann. Neurol., 9:34.

McFarlin, D. E., and McFarland, H. F., 1982, Multiple sclerosis, New Engl.J.Med., 309:1183.

Mehta, P. D., Lassmann, H., and Wisniewski, H. M., 1981, Immunologic studies of chronic relapsing EAE in guinea pigs - Similarities to multiple sclerosis, J.Immunol., 127:334.

Mehta, P. D., Mehta, S. P., and Patrick, B. A., 1982, Identification of light chain type bands in CSF and serum oligoclonal IgG from patients with multiple sclerosis, J.Neuroimmunol., 2:119.

Mehta, P. D., Patrick, B. A., and Wisniewski, H. M., 1981, Isoelectric focusing and immunofixation of cerebrospinal fluid and serum in multiple sclerosis (MS), J.Clin.Lab.Immunol., 6:17.

Mehta, P. D., Patrick, B. A., and Wisniewski, H. M., 1983, Oligoclonal IgG bands in chronic relapsing EAE and MS. Abstract from the Satellite Meeting of the International Society of Neurochem., Seattle, Washington.

Merrill, J., Wahlin, B., Siden, A., and Perlmann, P., 1982, Elevated direct and IgM enhanced ADCC activity in multiple sclerosis patients, J.Immunol., 128:1728.

Mokhtarian, F., McFarlin, D. E., and Raine, C. S., 1983, Induction of chronic relapsing experimental allergic encephalomyelitis by adoptive transfer of lymph node cells, J.Neuropath.Exp. Neurol., 42:359 (abstr).

Neuwelt, E. A., and Clark, W. K., 1978, "Clinical Aspects of Neuroimmunology," Williams and Wilkins Co., Baltimore.

Olsson, T., Henriksson, A., Link, M., and Kristensson, K., 1983, Humoral immune response in chronic relapsing experimental allergic encephalomyelitis in guinea pigs. Abstract from the Satellite Meeting of the International Society of Neurochem., Seattle, Washington.

Olsson, T., Kristensson, K., Leijon, G., and Link, H., 1982, Demonstration of serum IgG antibodies against myeling during the course of relapsing experimental allergic encephalomyelitis in guinea pigs, J.Neurol.Sci., 54:359.

Olsson, J., and Pettersson, B., 1975, A comparison between agar gel electrophoresis and CSF serum quotients of IgG and albumin in neurological diseases, Acta Neurol.Scandinav., 53:308.

Omechein, M., 1983, Inflammatory cells in the central nervous system: An integrated concept based on recent research in pathology, immunology and forensic medicine, in: "Progress in Neuropathology," H. M. Zimmerman, ed., 5:277.

Oppenheimer, D. R., 1978, The cervical cord in multiple sclerosis, Neuropath.Appl.Neurobiol., 4:151.

Paterson, P. Y., and Day, E., 1979, Neuroimmunologic disease: Experimental and clinical aspects, Hosp.Pract., 14:49.

Paterson, P. Y., and Whitacre, C. C., 1981, The enigma of oligoclonal immunoglobulin G in cerebrospinal fluid from multiple sclerosis patients, Immunol.Today 2:111.

Reinherz, E., Weiner, H., Hauser, S., Cohen, J., Distaso, J., and Schlossman, S., 1980, Loss of suppressor T cells in active multiple sclerosis: Analysis with monoclonal antibodies, N.Engl.J.Med., 303:125.

Rivers, T. M., Sprint, D. H., and Berry, G. P., 1935, Encephalomyelitis accompanied by myelin destruction experimentally produced in monkeys, J.Exp.Med., 61:689.

Roboz, E., and Henderson, N., 1959, Preparation and properties of water-soluble proteins from bovine cord with "allergic" encephalomyelitic activity, in: "Allergic Encephalomyelitis," M. W. Kies and E. C. Alvord, Jr., eds., Charles C. Thomas, Springfield, Illinois, pp.281-292.

Ryberg, B., 1978, Multiple specification of antibrain antibodies in multiple sclerosis and chronic meylopathy, J.Neurol.Sci., 38:357.

Salmi, A., Rennanen, M., Ilonen, J., Panelius, M., 1983, Intrathecal antibody synthesis to virus antigens in multiple sclerosis, Clin.Exp.Immunol., 52:241.

Salt, I., Westall, F. C., Romine, J. S., and Wiederholt, W. C., 1980, Studies on myelin and preliminary report on immunologic observations, in: "Progress in Multiple Sclerosis Research," H.J. Bauer and S. Poser, eds., Springer, Berlin, p. 419.

Schuller-Levis, G., Clausen, J., Schwerer, B., Madrid, R. E., and Wisniewski, H. M., 1983, Dynamics of cellular immunity during CR-EAE in guinea pigs. Abstract from the Satellite Meeting of the International Society of Neurochem., Seattle, Washington.

Schwerer, B., Egghart, M., Kitz, K., Lassmann, H., and Bernheimer, H., 1983, Serum antibodies against central nervous system (CNS) antigens in chronic relapsing experimental allergic encephalomyelitis (A-EAE), in: "Protides of the Biological Fluids," H. Peeters, ed., Pergamon Press, Oxford, pp.193-196.

Schwerer, B., Schuller-Levis, G. B., Mehta, P. D., Madrid, R. E., and Wisniewski, H. M., 1983, Cellular and humoral immune response to MBP during the course of chronic relapsing EAE. Abstract from the Satellate Meeting of the International Society of Neurochem., Seattle, Washington.

Seitelberger, F., 1967, Autoimmunologische aspekte der entmarkungsenzephalitiden, Nervenarzt 38:525.

Shaw, C. M., Alvord, E. C., Kaku, J., and Kies, M. W., 1965, Correlation of experimental allergic encephalomyelitis with delayed-type skin sensitivity to specific homologous encephalitogen, Ann.N.Y.Acad.Sci., 122:318.

Shevach, E., and Rosenthal, A., 1973, Function of macrophages in recognition by guinea pig T-lymphocytes. II. Role of the macrophages in the regulation of genetic control of the immune response, J.Exptl.Med., 138:1213.

Sobel, R. A., Blanchette, B. W., and Colvin, R. B., 1983, Pre-inflammatory expression of fibronectin (Fn) and Ia in acute experimental allergic encephalomyelitis (EAE): An active role for endothelial cells (EC) in the immune response detected by quantitative immunoperoxidase studies using monoclonal antibodies (MAb). Abstract from the Satellite Meeting of the International Society of Neurochem., Seattle, Washington.

Stone, S. H., and Lerner, E. M., 1965, Chronic disseminated allergic encephalomyelitis in guinea pigs, Ann.N.Y.Acad.Sci., 122:227.

Suckling, A. J., Reiber, H., Kirby, J. A., and Humsby, M. G., 1983, Chronic relapsing experimental encephalomyelitis: Immunological and blood-cerebrospinal fluid - barrier dependent changes in the CSF, J.Neuroimmunol., 4:35.

Tabira, T., Stoyama, Y., and Kuroiwa, Y., 1983, Is continuous antigenic stimulation at the immunized loci necessary for chronic relapsing EAE? Abstract from the Satellite Meeting of the International Society of Neurochem., Seattle, Washington.

Tourtellotte, W. W., 1976, Interaction of local central nervous system immunity and systemic immunity in the spread of multiple sclerosis demyelination, in: "Multiple Sclerosis: Immunology, Virology and Ultrastructure," G. Wolfgang, G. W. Ellison, J. G. Stevens, J. M. Andrews, eds., Academic Press, New York, pp.285-332.

Traugott, U., Stone, S., and Raine, C. S., 1978, Experimental allergic encephalomyelitis - Migration of early T cells from the circulation into the central nervous system, J.Neurol. Sci., 36:55.

Trotter, J., and Smith, M. E., 1983, Macrophage-mediated demyelination: the role of lipases and antibody. Abstract from the Satellite Meeting of the International Society of Neurochemistry, Seattle, Washington.

Uchimura, I., and Shiraki, H., 1957, A contribution to the classification and the pathogenesis of demyelinating encephalomyelitis, J.Neuropath.Exp.Neurol., 16:139.

Vartdal, F., Vandvik, B., and Norrby, E., 1980, Viral and bacterial antibody response in multiple sclerosis, Ann.Neurol., 8:248.

Voss, K., Lassmann H., Kitz K., Wisniewski, H. M., and Iqbal, K., 1983, Ultracytochemical distribution of myelin basic protein after injection into the cerebrospinal fluid: Evidence for transport through the blood brain barrier and binding to the luminal surface of cerebral veins, J.Neurol.Sci., in press.

Waksman, B., 1981, Current trends in multiple sclerosis research, Immunol.Today 2:87.

Waksman, B. H., and Morrison, L. R., 1951, Tuberculin-type sensitivity to spinal cord antigen in rabbits with isoallergic encephalomyelitis, J.Immunol., 66:421.

Weiner, H. L., and Hauser, S. L., 1982, Neuroimmunology I: Immunoregulation in neurological disease, Ann.Neurol., 11:437.

Wekerle, H., Cohen, I. R., and Feldman, H., 1974, Selbst-Toleranz und Autoimmunitat, Dtsch.med.Wschr., 99:1734.

Wisniewski, H. M., 1977, Immunopathology of demyelination in autoimmune diseases and virus infection, Br.Med.Bull., 33:54.

Wisniewski, H. M., Brown, H. R., and Thormar, H., 1983a, Pathogenesis of viral encephalitis: Demonstration of viral antigen(s) in the brain endothelium, Acta Neuropathol. 60:107.

Wisniewski, H. M., and Keith, A. B., 1977, Chronic relapsing EAE an experimental model of multiple sclerosis, Ann.Neurol., 1:144.

Wisniewski, H. M., and Lassmann, H., 1983b, Etiology and pathogenesis of monophasic and relapsing inflammatory demyelination - human and experimental, Acta Neuropathol., Suppl.IX., pp. 21-29.

Wisniewski, H. M., Lassmann, H., Brosnan, C. F., Mehta, P. D., Lidsky, A. A., and Madrid, R. E., 1982, Multiple sclerosis: Immunological and experimental aspects, in: "Recent Advances in Clinical Neurology," W. B. Matthews and G. H. Glaser, eds, Churchill Livingstone, Edinburgh, No. 3, pp. 95-125.

Wisniewski, H. M., and Madrid, R. E., 1983, Chronic progressive experimental allergic encephalomyelitis in adult guinea pigs, J.Neuropath.Exp.Neurol., 42:243.

Yung, L. L. L., Diener, E., McPherson, T. A., Barton, M. A., and Hyde, H. A., 1973, Antigen-binding by lymphocytes in normal man and guinea pig to human encephalitogenic protein, J.Immunol., 110:1383.

COMPARATIVE NEUROPATHOLOGY AND PATHOGENETIC ASPECTS OF CHRONIC RELAPSING EXPERIMENTAL ENCEPHALOMYELITIS AND MULTIPLE SCLEROSIS

Franz Seitelberger and Hans Lassmann

Neurological Institute of the University of Vienna
Schwarzspanierstr. 17
A-1090 Vienna

Multiple sclerosis (MS) is a disease well known clinically for more than 200 years. The first detailed description of its pathology has been provided by Charcot in Paris in the late 19th Century. SInce that time many research efforts have been devoted to clarify the etiopathogenesis of MS and to find some effective treatment, yet so far without convincing success.

The disease MS had been named from its characteristic neuropathological finding, the sclerotic plaque. The plaque, however, already is a late product of the underlying pathological process, apparently representing the final deleterious stage of the clinical disease, manifested in the CNS. All steps of this final phase of the process are supposed to have their corresponding morphological correlates in the target organ, the CNS. Therefore it is surprising that since the classical neuropathological descriptions, which after Charcot were presented among others by Marburg in Vienna, Hallervorden in Berlin, to mention 2 of the most prominent workers, only very few original studies of MS have been performed. That is regrettable because in the past decades not only have several potent morphological methods been developed but also numerous relevant experimental studies on inflammatory demyelination were done. Thus, there is a certain deficit above all in comparative investigations of human and experimental demyelination and some stagnation of disease-oriented basic research in this important field of neurology. Due to the recent development of reproducible chronic models of EAE (Stone and Lerner 1965, Wisniewski and Keith 1977, Massanari 1980, Lublin 1982) a new challenging situation arose: The far reaching conformity of the clinical and pathological findings in cr-EAE and MS makes it possible to draw exact comparisons between the series of tissue events, their quality and time course, and to identify the position of the individual changes related to the coordinates of the complex disease process.

Before discussing some selected results of our recent extensive comparative studies I would like to sketch the pathological aspects of human MS and its position among the demyelinating disease.

The neuropathological diagnosis of MS is based on the presence of focal lesions presenting two different alterations, namely chronic inflammatory changes associated with elective demyelination, i.e. loss of myelin sheaths without axonal damage. This essential combined lesion pattern of the foci is in itself variable concerning intensity, cellular composition of the perivenous and perivenular inflammation, size and topography of demyelination, and affection of oligodendroglia. These primary changes have also partial influence upon the orthological reactions of the central nervous tissue following the lesion especially regarding astroglial reparative proliferation with scar formation and remyelination depending on the preservation of oligodendrocytes. The state of a focal MS-lesion encountered will also depend on its individual age, determining the intensity of myelin degradation and inflammation but also of remyelination and scar formation. Furthermore it is of utmost importance that the lesions in MS are significantly different dependent on their development in different phases (early or late) of the disease along its chronic course: that stage-dependent quality of MS-foci must be strictly distinguished from the histological age of an individual plaque.

From the above findings it can be concluded that there is besides the variable dualistic structure of the fundamental histological alterations in every case of MS a large intraindividual variability of the lesions and between single cases a remarkable interindividual variability. These circumstances make it conceivable that several subtypes of MS, also considering clinical features, may exist:

1) typical relapsing - remittent
2) chronic progressive
3) subacute ponto-cerebellar (Redlich)
4) acute (Marburg)

There is furthermore the group of inflammatory demyelinating diseases among which MS is the most prominent but often not exactly distinguishable example.

1) acute hemmorrhagic leukoencephalitis
2) acute disseminated encephalomyelitis
3) inflammatory diffuse sclerosis (Schilder)
4) encephalitis concentrica (Balo)
5) neuromyelitis optica (Devic)
6) myelitis necroticans

Finally the differential diagnosis of MS against the monophasic postvaccinal and parainfectious diseases is sometimes difficult, particularly because transitional cases with MS are described in the literature.

We feel that all variations and overlappings, especially the inherent metamorphosis of the individual disease, have not been duly respected in the past and that therefore a rehearsal of MS neuropathology from the basis of the present knowledge is strictly indicated. Before that a short review of the neuropathology of cr-EAE will be given.

A good animal model of MS should fulfill the following criteria (Wisniewski et al., 1982): Clinically: Course in relapses and remissions, symptoms of multifocal central nervous lesions. Neurophysiologically: Oscillations of conduction velocity and amplitude. Neuropathologically: Primary inflammatory demyelination with foci in different stages. Neuroimmunologically: Oligoclonal bands of IgG in the CSF, IgG-synthesis in the CNS and participation of immunoregulatory cells.

As mentioned before cr-EAE appears to be the most close experimental model of MS available at present. The extensive neuropathological analysis of this model has demonstrated that it can simulate all stages and variable forms of human MS and all the different inflammatory demyelinating diseases of MS-type (Lassmann 1983). This potentiality depends above all upon the time interval between sensitization procedure and sampling of the animals.

In the following we will report on the most relevant findings of our neuropathological analysis of cr-EAE and then present some results of comparative study between this model and human MS.

In the experiment of cr-EAE the initial disease occurs 10-20 days after sensitization in the form of acute perivenous inflammation without demyelination or in severely affected animals in the form of an acute hemorrhagic leukoencephalitis (Lassmann and Wisniewski 1979a). There is not typical focal demyelination or plaque formation. Animals with active disease during the subacute stage of the disease (20-40 days after sensitization) show the classical picture of acute perivenous leukoencephalitis with extensive perivenous mononuclear cell infiltration and sleeve-like perivenous demyelination (Lassmann et al., 1980).

The first relapse with focal and substantial demyelination appears 40-60 days after sensitization. With the onset of initial clinical symptoms of the relapse in the respective area an invasion of mononuclear cells from the perivenous inflammatory spaces into the white substance can be seen (Lassmann and Wisniewski 1978). Many of these inflammatory cells are T-lymphocytes (Traugott et al., 1982).

Immediately after that the myelin sheaths are fragmented to so-called "myelin balls" and "myelin droplets" (Lassmann and Wisniewski 1979b). These myelin balls present histochemical qualities similar to the intact myelin sheaths concerning the myelin lipids, that means they are Luxol fast blue - and Marchi-positive. Myelin degradation in the fragments starts 10-14 days later with the appearance of sudanophilic, PAS-positive degradation products which like the myelin balls are taken up by phagocytes. Simultaneously also a glial reaction in the form of increase in size and number of astrocytes appears. The duration of the degradation phase depends on the size and localization of the focus. Superficial foci may be cleared of degradation debris already 14 days after the focus formation. In foci located in the depth of the white matter the degradation may last up to several months.

In the acute and early chronic stages CSF findings indicate a transient blood-brain barrier damage which correlates with disease activity. Intrathecal immunoglobulin synthesis is not apparent from the CSF protein profiles. The late chronic stage (100-400 days after sensitization) more closely resembles chronic MS. The inflammatory response in the CNS, although invariably present, is much less severe than in earlier stages, even in actively demyelinating lesions. Numerous demyelinated sclerotic or partially remyelinated lesions are found throughout the neuraxis in similar topographical distribution as in human MS. The majority of the lesions are inactive or repaired, only few actively demyelinating plaques may be found in animals sampled during a relapse of the disease. CSF protein shows a partial or complete repair of blood-brain-barrier function and evidence for intrathecal immunoglobulin synthesis (Mehta et al., 1981, Clynn et al., 1982, Karcher et al., 1982, Lassmann et al., 1983).

With regard to the modality of the myelin sheath decomposition the following processes can be distinguished:

1. Direct attack of phagocytes against the myelin sheath in the form of "myelin stripping" (Lampert 1965). This change mostly starts from the paranodal area; it is accompanied, however, by the disintegration of the total internodium.
2. Another process is the so-called "vesicular disruption of myelin" which is encountered in variable intensity (Lampert 1965, Dal Canto et al., 1975). This type seems to be a pathogenetically independent process.
3. The low incidence of myelin stripping and vesicular disruption of meylin in active cr-EAE plaques, however, indicates that other, less characteristic alterations of the myelin sheath may represent initial stages of demyelination too (Lassmann 1983).

Myelin degradation is unanimously thought to be performed by hematogenic macrophages (Kosunen et al., 1963, Lampert and Carpenter 1965). Local glial cells, however, may participate as one can find so-called "fixed degradation" in astro- and probably oligodendrocytes (lassmann 1983). This type of "Abbau" seems to be more prominent in foci of the chronic phase and less prominent in the acute phase of the disease.

With regard to the shape and formation of the demyelinating foci one can distinguish the perivenous demyelinating seams, typical plaques, plaques with peripheral (radial) extension zones and superficially located so-called "Mantel-Herde" (Lassmann 1983). These types of foci do not occur irregularly but according to the respective varying pathogenetic situation along the course of the experimental disease. The "Mantel-Herde", e.g. seem to be conditioned by the increased pathogenicity of CSF, probably related to extensive seeding of inflammatory cells in the meninges (Lassmann et al., (1981a) and on the ventricular surface (Lassmann et al., 1981b).

The distribution and topography of foci in cr-EAE has been studied in relation to the perivenous inflammatory changes in the pia arachnoid of spinal cord. It could be demonstrated that spot-like dense cuffs of hematogenic cells appear around the postcapillary radial venules draining the spinal cord: so-called "Stich-Venen". Respective areas of the spinal cord are predilection sites for demyelinated plaques (Lassmann et al., 1981a).

Role of Oligodendroglia

As the myelin sheaths sensu stricto are parts of the oligodendroglial cells, the possibility of a pathogenic effect to the meylin sheaths via the oligodendroglia has long been discussed. The solution of the problem is difficult because identification of oligodendroglia in the active demyelinating focus is not easy and often impossible even if using elective staining methods and electron microscopy. From the neuropathological standpoint, however, a primary affection of oligodendroglia is not probable because histological alterations which precede de-myelination cannot be seen. Furthermore the demyelination pattern in diseases due to primary affections of oligodendroglia, e. g. viral encephalitis with inclusion bodies and Progressive Multifocal Leukoencephalopathy (PML) differs essentially from the focal demyelination in EAE and in MS. Finally the satellite oligodendroglia cells in the grey matter are always completely unaffected.

In foci of the experimental disease the oligodendroglia are reduced to variable degrees. The highest rate of oligodendroglia loss is generally found in late chronic foci in which correspondingly remyelination is slight whereas in more acute foci loss of oligodendroglia cells is moderate and therefore an intense remyelination can be seen (Lassmann et al., 1980, Lassmann 1983). These foci correspond to the so-called shadow-plaques which therefore are not foci of incomplete demyelination but represent foci with intensive remyelination. This may possibly be due to an immunopathological cause based on the increase of antimyelin antibodies during the chronic stage of the disease (Schwerer et al., 1983, Lassmann et al., 1983). Such antibodies may not only be involved in demyelination and oligodendroglia destruction (Bornstein and Appel 1961, Raine and Bornstein 1970, Lassmann et al., 1981c) but may also inhibit remyelination (Seil et al., 1973, Bornstein and Raine 1970). It is this phase of the disease which produces the typical myelin-free sclerotic plaques rich in glial fibers. There occurs also the peripheral type of remyelination by invading Schwann cells from the spinal nerve roots into demyelinating plaques of the spinal cord (Raine et al., 1978).

MULTIPLE SCLEROSIS

The Inflammatory reaction

In multiple sclerosis (MS) perivenous inflammatory infiltrates composed of lymphocytes, plasma cells, large mononuclear cells and macrophages are regularly present (Marburg 1936). In an extensive study on autopsy cases of MS (Guseo and Jellinger 1975) inflammatory infiltrates were found in 60% of total cases and in 74% of actively demyelinating cases. MS cases without perivenous inflammation were either inactive (burned out) or were pre-treated with immunosuppressive therapy. Thus from neuropathology there is so far no hard evidence for the occurrence of actively demyelinating lesions in the absence of inflammation although it cannot be definitely excluded in chronic MS. It is, however, even more difficult to prove the primary nature of inflammation, i.e. that inflammation always precedes demyelination and is not a secondary alteration due to tissue damage. The main support for the primary nature of inflammation in MS comes from the study of acute MS cases. In these cases a continuous transition of the classical perivenous inflammatory lesions of acute disseminated luekoencephalitis type into typical MS plaques is regularly found (Marburg 1904, Pette 1928, Seitelberger 1973, Krücke 1973). The intensity of the inflammatory reaction is very high in acute MS whereas in chronic MS inflammation is much less pronounced even in active plaques. The number of inflammatory cells, however, in chronic inactive MS cases is still rather high (Prineas and Wright 1978). The number of inflammatory cells withing chronic MS plaques varies from case to case and even from plaque to plaque within the same case (Guseo and Jellinger 1975). Concerning distribution of inflammatory cells in acute MS the infiltrates are arranged around all segments of the venous vascular tree (Seitelberger 1973). In chronic MS inflammation is focal, predominantly affecting large drainage veins (Seitelberger 1973). The inflammatory cells are generally localized in a perivenous position. In the demyelinating lesions lymphoctyes and plasma cells are present only in small numbers. The high cellularity of active MS plaques is mainly due to phagocytes (Hallervorden 1940, Guseo and Jellinger 1975. The presence of T-cells in MS plaques has been shown histochemically (Traugott et al., 1983).

Vascular Pathology

Compared with cr-EAE vascular pathology in MS is not as prominent. It is still unresolved, however, to what extent occlusion or damage to small vessels which sometimes occurs in MS plaques (Siemerling and Raeke 1914) may contribute to the lesional growth (Putnam 1935). The intensity of chronic vascular changes like perivenous and pericapillary fibrosis is variable from case to case. These alterations are found mainly in the same areas where pericapillary accumulation of phagocytes is pronounced in active lesions (Lassmann 1983).

Demyelination and Myelin Degradation

Elective demyelination is the most characteristic event in the pathology of MS. In MS the identification of initial structural changes leading to demyelination is rather difficult. The most frequent alteration of myelin in lesions which were believed to be active is splitting of myelin lamellae and the formation of small intramyelinic vesicles (Perier and Gregoire 1965, Suzuki et al., 1969, de Preux and Mair 1974, Kirk 1979). Recently also a lesion called micropinocytosis vermiformis (Prineas and Connell 1978) has been described as the initial alteration. It should, however, be stressed that the presence of sudanophilic myelin degradation products used as a marker of lesional activity in the above mentioned studies is not a good marker for initial lesions. Based on the experience of the study of the sequences of events in cr-EAE, the pattern of demyelination in cases of acute and chronic MS from the collection of the Neurological Institute has recently been investigated (Lassmann 1983). In acute MS the most frequent initial alteration in all investigated lesions was extensive vesicular disruption of myelin. This transformation either involved the whole myelin sheath or was localized in the inner or outer loops of the sheath, especially near the node of Ranvier. In the same areas invasion of phagocytes or their cell processes into the myelin sheath was frequently noted corresponding to myelin stripping. Intramyelinic phagocytes were found either between the axon and the myelin sheath or between myelin lamellae. In the vicinity of completely or partially demyelinated fibers numerous phagocytes were observed containing fragments of the myelin sheaths. Active plaques in chronic MS differed from the demyelinating lesions in acute MS in so far as the intensity of perivenous inflammatory reaction together with the hypercellularity of actively demyelinating areas was less pronounced and the invasion of phagocytes into the myelin sheaths was rare.

Initial stages of demyelination in these cases included partial vesicular disruption of myelin sheaths. The main structural change, however, consisted of intramyelinic edema with massive swelling of the sheaths, widening of the node of Ranvier and paranodal asymmetrical thinning and removal of myelin sheaths.

Further myelin degradation in MS follows a standard histological, histochemical and ultrastructural pattern corresponding to that found in other inflammatory demyelinating diseases or that of diseases with secondary destruction of myelin sheaths (Seitelberger 1960). Myelin degradation starts with a stage of physical disintegration of the myelin sheath followed by the chemical degradation of myelin lipids, resulting in the appearance of sudanophilic and later of PAS-positive material. Sudanophilic lipids seem to be present within the lesions up to several months after plaque formation. The abundance of PAS-positive ultrastructurally pleomorphic degradation products (Prineas 1975) cannot be taken as characteristic for early MS plaques but is also found in other demyelinating conditions (Seitelberger 1960, Hauw and Escourolle 1977, Lassmann et al., 1978). The accumulation of debris-containing phagocytes in the pericapillary region is more pronounced in acute MS than in active lesions of chronic MS. A proportion of debris is digested in local cells such as astroglia (Marburg 1906).

Thus it can be stated that the patterns of demyelination and myelin degradation are found to be very similar in cr-EAE and MS, provided active lesions are identified in MS on the basis of myelin degradation products.

Oligodendroglia

The interpretation of oligodendrocyte alterations in MS is of special importance for the pathogenetic interpretation of demyelinating lesions and especially regarding the question of whether the disease process is primarily directed against myelin or whether demyelination follows alterations or destruction of oligodendrocytes. In the literature there are divergent opinions about the fate of oligodendrocytes in MS (McAlpine et al., 1955, Ibrahim and Adams 1963, Lumsden 1970, Itoyama et al., 1980, Raine et al., 1981). Our own studies in accordance to the results obtained by Ibrahim and Adams (1963, 1965) indicate that there are differences in the extent of oligodendroglia destruction between small lesions of acute and rapidly progressive chronic MS compared with the typical large plaques of chronic MS (Lassmann 1983). In a general pattern oligodendrocytes are most effectively destroyed in the large lesions of chronic MS although even in that situation a plaque completely free of oligodendroglia-like cells is rare. In acute and rapidly progressive chronic MS cells resembling oligodendroglia may be present in active completely demyelinated plaques in numbers comparable to those in normal periplaque white matter. Furthermore remyelination in these plaques is pronounced, occurring within a few days to weeks after formation of the lesion. Therefore MS does not seem to be a primary disorder of oligodendrocytes. This is further supported by the pattern of demyelination reported above suggesting that myelin is the main target in the pathogenesis of the disease. Oligodendrocytes as a functional unit with myelin sheaths may degenerate to a variable extent during active demyelination. Alternatively the immune reaction in different MS patients may be variable, directed either against myelin antigens alone or in other cases against antigens shared by myelin and oligodendroglia (Lassmann 1983).

Remyelination

Due to the difficulty in the identification of remyelination in tissue obtained from autopsy, this important repair mechanism had not been recognized and respected enough in the past. In this connection it is interesting that already Marburg (1906) in a detailed and sequential study of plaque formation in acute MS had suggested that shadow plaques in MS can be explained on the basis of remyelination. In recent electron microscopic studies the presence of remyelination at the margins of MS plaques has been documented (Suzuki et al., 1969, Prineas and Connell 1979). Remyelination as a repair mechanism of whole plaques, however, has been questioned up to now. The study of the dynamics of remyelination in cr-EAE, however, helped to understand the factors regulating the extent of remyelination also in the human disease. The phenomenon of remyelination in MS is mostly connected with the occurrence of so-called shadow plaques. The occurrence of these had been explained by a concept of incomplete demyelination (Lumsden 1970). At present it is, however, indicated from our results that the responsible mechanisms of the appearance of shadow plaques is remyelination (Lassmann 1983). In acute MS shadow lesions are especially frequent. In chronic MS cases shadow plaques although present in variable numbers are rare compared with acute MS. From studies of chronic MS cases with short duration the sequence of events leading to the formation of shadow plaques is the following: Already during the sudanophilic stage of myelin degradation, simultaneously with the onset of fibrillary sclerosis, small clusters of thinly myelinated nerve fibres appear. Later, areas of myelin pallor with uniform distribution of thinly myelinated fibers are present. Shadow plaques finally are characterized by a decreased density of myelinated fibers and by myelin sheaths which are still thinner than those of the surrounding white matter. The presence of sclerotic plaques without detectable myelin loss in some MS cases indicates that even complete remyelination may occur. Thus in general lesions in cases with short clinical history, especially acute MS cases, remyelinate more effectively than chronic cases. Remyelination may start rapidly after plaque formation. With chronicity of the disease not only the extent but also the velocity of remyelination decreases in newly formed lesions. The rapid and complete clinical recovery of patients after the first exacerbations of the disease may be partly due to rapid and extensive remyelination.

Sclerosis

The formation of a dense glial scar in MS is one of the most characteristic aspects of its pathology. Glial scar formation is a general repair mechanism in the central nervous system. There had been, however, a discussion whether sclerosis in MS is an event involved in the pathogenesis of the disease or whether it merely represents a sequel of tissue destruction (Müller 1904, Prinease and Raine 1976, McKeown and Allen 1978). Studies especially from acute MS demonstrated that sclerosis is clearly a secondary event following demyelination and was not found earlier than 14 days after plaque formation (Pette 1928). The first change is intense proliferation and increase in size of astrocytes with often bizarre size and forms and several nuclei (Peters 1935). In active lesions of chronic MS cases, however, sclerosis was found to be a very early feature in plaque formation and frequently simultaneously with active demyelination. It should be mentioned that there is some indication that inflammatory pathogenetic factors such as lymphocyte products or liberated debris may activate astrocytes and thus influence the density of glial scar formation (Fontana et al., 1980). On the other hand glia activation may possibly contribute to the pathogenesis of the demyelinating lesions via proteolytic enzymes (McKeown and Allen 1978).

Neuronal and Axonal Pathology

It is well known that in MS reduction of axons occur and nerve cell changes can be found which in general are interpreted as secondary alterations due to tissue damage. In MS the degree of axonal loss is variable from plaque to plaque also within a single case. In most cases axonal loss appears to correspond to the intensity of the inflammatory reaction and demyelination during the active phase of the lesions. However, considering exceptional cases with severe axonal and neuronal damage it cannot be excluded that in some cases the immune reaction may be directed not only against myelin/oligodendrocytes but also against nerve cell antigens (Fraenkel and Jakob 1913).

Peripheral Nervous System Pathology

MS is generally believed to be an exclusive central nervous system disease (Lumsden 1970). However, some cases have been described of indisputable plaque-like inflammatory demyelination in the

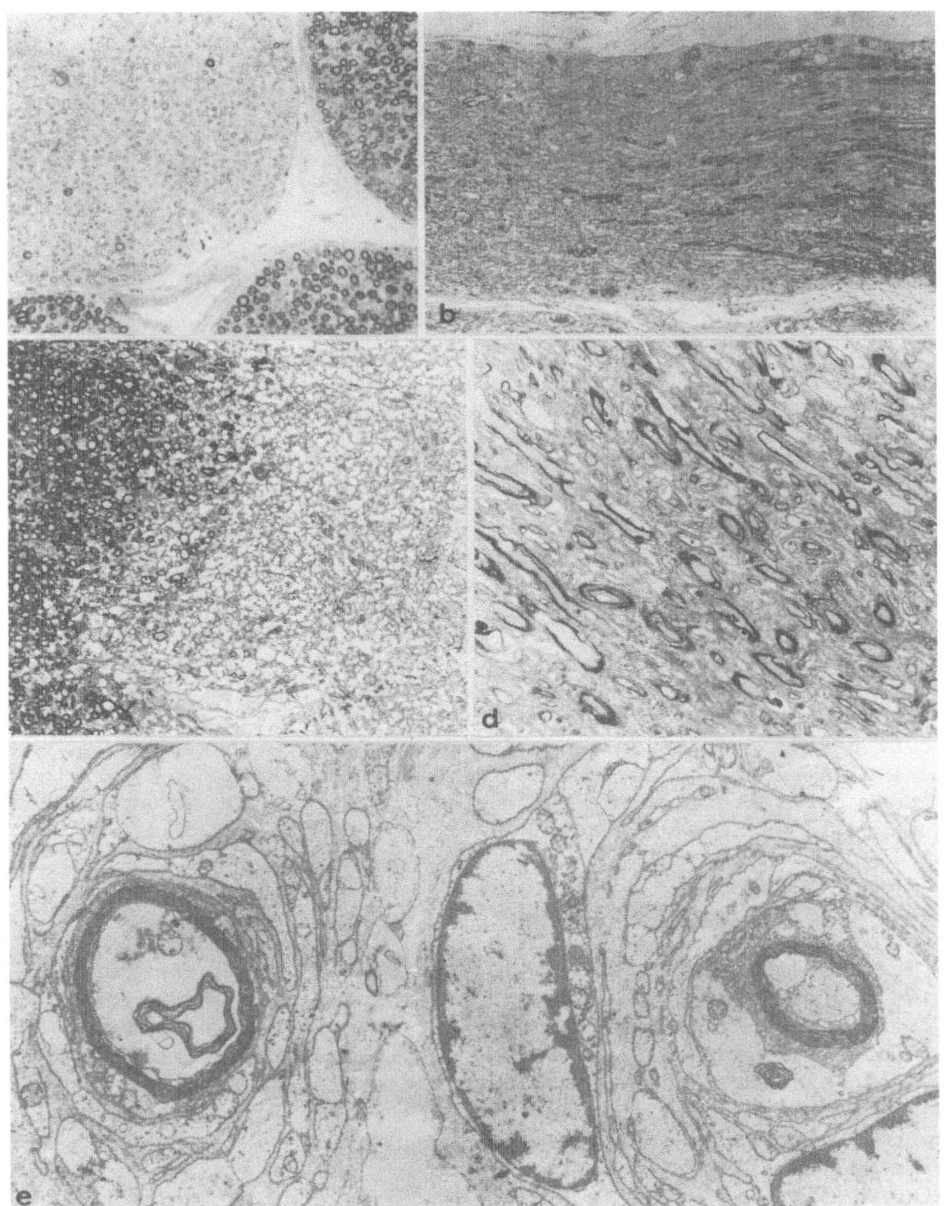

Fig. 1. a,b: 26 years old woman; acute MS; clinical history and neuropathological findings published elsewhere (Lassmann et al., 1981d). a: Nerve root of cauda equina; thin myelin sheaths representing remyelination. Toluidine blue, x 150. b: Cervical spinal root; plaque-like demyelination and remyelination in the peripheral portion of the root.

Fig. 1. Continued

Toluidine blue, x 150. c,d,e: 34 years old man: chronic MS. c: Border zone of an old plaque in the optic nerve. d: Olfactory nerve: decreased density of myelinated fibers; onion bulbs. Toluidine blue, x 630. e: Olfactory nerve; onion bulb formation. EM x 4 000.

peripheral nerves and roots, although the incidence of peripheral nervous system (PNS) involvement in classical chronic MS seems to be very low (Jellinger 1969). Of course changes like secondary nerve damage due to malnutrition, vitamin deficiency etc. have to be excluded. Inflammatory demyelination in the PNS, however, is frequent in Marburg's type of acute MS (Marburg 1906, Pette 1928, Lassmann et al., 1981d). As in EAE demyelination has been noted mainly in the spinal roots. The spectrum of alterations include inflammation, primary demyelination, plaque-like distribution of the lesions, remyelination and the formation of onion bulbs. The factors responsible for the cross reaction between CNS and PNS myelin in human diseases are not yet known.

Mechanisms of Plaque Growth

In cases of acute MS the main type found is the confluence of adjacent perivenous lesions. Radial plaque growth is generally found in active cases of chronic MS. It must be emphasized, however, that confluent perivenous lesions may also occur in chronic MS and radial plaque growth in acute MS cases.

Lesional Topography in the CNS

The distribution of lesions in the CNS has been intensively studied for may years. The pattern of demyelination and lesional distribution varies during chronicity of the disease. In acute perivenous leukoecephalitis perivenous inflammatory infiltrates and perivenous sleeves of demyelination are disseminated and may be found at any location in the CNS (Seitelberger 1973). In contrast in chronic MS demyelinated plaques are focal, oriented towards larger vessels and localized mainly in certain predilection sites in the CNS (Steiner 1931), Seitelberger 1973). In chronic MS the relationship of large demyelinated plaques to the distribution of large veins is clearly established and is especially well documented in the spinal cord (Fog 1950). Other factors e.g. the mechanical strain on the vasculature may help to precipitate a lesion at a given place (Oppenheimer 1978). In addition there is another plaque type in chronic Ms which is related to the inner and outer surfaces of CNS. The most classical lesion of this type is the so-called "Mantel-Herd" in chronic MS (Steiner 1931). Other predilection sites are the lateral angles of the lateral ventricles, the so-called "Wetterwinkel", the cerebellar peduncles, the cortical-subcortical areas, the lateral surface of the spinal cord and the optic nerves. In acute MS, however, periventricular lesions are rare and, if present, small and not related to the ventricular surface. It may be interesting that in acute MS cases the predilection sites for lesions in typical chronic MS are apparently not more frequently involved than other regions of the brain. It should be noted that in MS there is an inverse relationship between involvement of the brain and spinal cord which is best exemplified in cases of spinal MS and Devic's disease, where the involvement of the brain is minimal with the exception of the optic nerves (Seitelberger 1982).

Many of these pathohistological aspects of MS have been extensively described and well documented during the past 100 years. It was, however, difficult to interpret these individual aspects of MS pathology in their temporal and sequential order of plaque development. The close similarity of structural aspects of inflammatory demyelinating lesions in cr-EAE and MS, however, has helped to rearrange the structural details of demyelinating lesions in MS into a sequential concept of plaque formation.

Furthermore the comparative study of the neuropathology of cr-EAE and MS has led to a different conceptual approach in the interpretation of the pathology in these diseases. Although being aware of the broad spectrum of inter- and intraindividual variability of the structure of MS-plaques, most neuropathologists primarily tried to extrapolate the features of the lesions common to all MS-cases. Pathohistological and pathogenetic studies in cr-EAE, however, indicate that multiple immunopathogenetic mechanisms directed against multiple CNS-antigens, different from animal to animal, are involved in the formation of the lesions (Wisniewski et al., 1982, Lassmann 1983, Wisniewski and Lassmann 1983). It thus appears especially necessary to reemphasize a fact known in neuropathology since the beginning of this disease: Besides the obligatory alterations in MS (i.e. chronic perivenous inflammation, plaque-like demyelination and reactive gliosis) involvement of other structures of the CNS is regularly found although variable from case to case and from lesion to lesion. This, however, at the same time is a warning not to draw far reaching pathogenetic conclusions from sophisticated studies on a single MS-plaque or a single MS-case.

REFERENCES

Bornstein, M. B., and Appel, S. H., 1961, The application of tissue culture to the study of experimental "allergic" encephalomyelitis. I. Patterns of demyelination, J. Neuropathol.Exp. Neurol., 20:141.

Bornstein, M. B., and Raine, C. S., 1970, Experimental allergic encephalomyelitis: antiserum inhibition of myelination in vitro, Lab.Invest., 23:536.

Charcot, J. M., 1868, Histologie de la sclerose en plaque, Gaz Hopital, 41:554.

Dal Canto, M. C., Wisniewski, H. M., and Johnson, A. B., 1975, Vesicular disruption of myelin in autoimmune demyelination, J.neurol.Sci., 24:313.

De Preux, J., and Mair, W. G. P., 1974, Ultrastructure of optic nerve in Schilder's disease, Devic's disease and disseminated sclerosis, Acta Neuropath., 30:225.

Fog, T., 1950, Topographic distribution of plaques in the spinal cord in multiple sclerosis, Arch.Neurol., 63:382.

Fontana, A., Grieder, A., and Arrenbrecht, St., 1980, In vitro stimulation of glia cells by a lymphocyte produced factor, J.neurol.Sci., 46:55.

Fraenkel, M., and Jakob, A., 1913, Zur Pathologie der multiplen Sklerose mit besonderer Berücksichtigung der akuten Formen, Z.Neurol., 14:565.

Glynn, P., Weedon, D., and Edwards, J., 1982, Humoral immunity in chronic relapsing experimental allergic encephalomyelitis. The major oligoclonal bands are antibodies to mycobacteria, J.neurol.Sci., 57:369.

Guseo, A., and Jellinger, K., 1975, The significance of perivascular nfiltrations in multiple sclerosis, Neurol., 211:51.

Hallervorden, J., 1940, Die zentralen Entmarkungs-erkrankungen, tsch.Z.Nervenheilk., 150:201.

Hauw, J. J., and Escourolle, R., 1977, Filaments and multilamellated cytoplasmic inclusions in progressive multifocal leukoencephalopathy, Acta Neuropath., 37:263.

Ibrahim, M. Z. M., and Adams, C. W. M., 1963, The relationship between enzyme activity and neuroglia in plaques of multiple sclerosis, J.Neurol.Neurosurg.Psychiat., 26:101.

Ibrahim, M. Z. M., and Adams, C. W. M., 1965, The relationship between enzyme activity and neuroglia in early plaques of multiple sclerosis, J.Path.Bacteriol., 90:239.

Itoyama, Y., Sternberger, N. H., and Webster, H. de F., 1980, Immunocytochemical observations on the distribution of myelin associated glycoprotein and myelin basic protein in multiple sclerosis lesions, Ann.Neurol., 7:167.

Jellinger, K., 1969, Einige morphologische Aspekte der multiplen Sklerose, Wien.Z.Nervenheilk., Suppl.II:12.

Karcher, D., Lassmann, H., and Lowenthal, A., 1982, Antibodies-restricted heterogeneity in serum and cerebrospinal fluid of chronic relapsing experimental allergic encephalomyelitis, J.Neuroimmunol., 2:93.

Kirk, J., 1979, The fine structure of the CNS in multiple sclerosis. II. Vesicular demyelination in a acute case, Neuropathol.Appl. Neurobiol., 5:289.

Kosunen, T. V., Waksman, B. H., and Samuelsson, K., 1963, Radioautographic study of cellular mechanisms in delayed hypersensitivity.II. Experimental allergic encephalomyelitis in the rat, J.Neuropathol.Exp.Neurol., 22:367.

Krücke, W. 1973, On the histopathology of acute hemorrhagic leucoencephalitis, acute disseminated encephalitis and concentric sclerosis, Int. Symposium on aetiology and pathogenesis of the demyelinating diseases, Kyoto, pp.11.

Lampert, P. W., 1965, Demyelination and remyelination in experimental allergic encephalomyelitis, J.Neuropathol.Exp.Neurol., 24:371.

Lampert, P. W., and Carpenter, S., 1965, Electron microscopic studies on the vascular permeability and the mechanisms of demyelination in experimental allergic encephalomyelitis, J.Neuropathol.Exp.Neurol., 24:11.

Lassmann, H., 1983, Comparative neuropathology of chronic experimental allergic encephalomyelitis and multiple sclerosis, Neurology Series Vol.25, Springer, Berlin, Heidelberg, New York, Tokyo.

Lassmann, H., and Wisniewski, H. M., 1978, Chronic relapsing EAE. Time course of neurological symptoms and pathology, Acta Neuropathol., 43:35.

Lassmann, H., and Wisniewski, H. M., 1979a, Chronic relapsing experimental allergic encephalomyelitis. Effect of age at the time of sensitization on clinical course and pathology, Acta Neuropathol., 47:111.

Lassmann, H., and Wisniewski, H. M., 1979b, Chronic relapsing experimental allergic encephalomyelitis: Morphological sequence of myelin degradation, Brain Res., 169:357.

Lassmann, H., Ammerer, H. P., and Kulnig, W., 1978, Ultrastructural sequence of myelin degradation. I. Wallerian degeneration of the rat optic nerve, Acta Neuropathol., 44:91.

Lassmann, H., Kitz, K., and Wisniewski, H. M., 1980, Structural variability of demyelinating lesions in different models of subacute and chronic experimental allergic encephalomyelitis, Acta Neuropathol., 51:191.

Lassmann, H., Kitz, K., and Wisniewski, H. M., 1981a, Histogenesis of demyelinating lesions in the spinal cord of guinea pigs with chronic relapsing experimental allergic encephalomyelitis, J.neurol.Sci., 50:109.

Lassmann, H., Kitz, K., and Wisniewski, H. M., 1981b, The development of periventricular lesions in chronic relapsing experimental allergic encephalomyelitis, Neuropathol.Appl.Neurobiol., 7:1.

Lassmann, H., Kitz, K., and Wisniewski, H. M., 1981c, In vivo effect of sera from animals with chronic relapsing experimental allergic encephalomyelitis on central and peripheral myelin, Acta Neuropathol., 55:297.

Lassmann, H., Budka, H., and Schnaberth, G., 1981d, Inflammatory demyelinating polyradiculitis in a patient with multiple sclerosis, Arch.Neurol., 38:99.

Lassmann, H., Karcher, D., Kitz, K., Lowenthal, A., and Verslegers, W., 1983a, Humoral immune response in guinea pigs of different stages of EAE, in: "Experimental Allergic Encephalomyelitis: A Good Model for Multiple Sclerosis," E. C. Alvord, and M. W. Kies, eds., A. Liss Inc. New York (in press).

Lassmann, H., Suchanek, G., Kitz, K., Stemberger, H., Schweret, B., and Bernheimer, H., 1983b, Antibodies in the pathogenesis, of demyelination in chronic relapsing EAE, in: "Experimental Allergic Encephalomyelitis: A Good Model for Multiple Sclerosis," E. C. Alvord, M. W. Kies, and A. J. Suckling, eds., A. Liss. Inc. New York, (in press).

Lublin, F. D., 1982, Delayed relapsing experimental allergic encephalomyelitis in mice, J.neurol.Sci., 57:105.

Lumsden, C. E., 1979, The neuropathology of multiple sclerosis, in: "Handbook of Clinical Neurology," P. I. Vinken, and G. W. Bruyn, eds., Elsevier, New York, vol.9, p.217.

Marburg, O., 1906, Die sogenannte "akute Multiple Sklerose," ahrb.Psychiatrie, 27:211.

Marburg, O., 1936, Multiple Sklerose, in: "Handbuch der Neurologie," O. Bumke, and O. Foerster, eds., Springer, Berlin, vol. 13/12,

Massanari, R. M., 1980, A latent relapsing neuroautoimmune disease in syrian hamsters, Clin.Immunol.Immunopathol., 16:211.

McAlpine, D., Compston, N. D., and Lumsden, C. E., 1955, "Multiple Sclerosis," Livingston, Edinburgh.

McKeown, S. R., and Allen, I. V., 1978, The cellular origin of lysosomal enzymes in the plaque in multiple sclerosis: a combined histological and histochemical study, Neuropath.Appl.Neurobiol., 4:471.

Mehta, P. D., Lassmann, H., and Wisniewski, H. M., 1981, Immunological studies of chronic relapsing EAE in guinea pigs: similarities to multiple sclerosis, J.Immunol., 127:334.

Müller, E., 1904, "Die multiple Sklerose des Gehirns und Rückenmarks," Fischer, Jena.

Oppenheimer, D. R., 1978, The cervical cord in multiple sclerosis, Neuropathol.Appl.Neurobiol., 4:151.

Perier, O., and Gregoire, A., 1965, Electron microscopic features of multiple sclerosis lesions, Brain, 88:937.

Peters, G., 1935, Zur Frage der Beziehungen zwischen der disseminierten nicht eitrigen Enzephalomyelitis und der multiplen Sklerose, Z.gesamt Neurol.Psychiatr., 153:356.

Pette, H., 1928, Über die Pathogenese der multiplen Sklerose, Dtsch.Z.Nervenheilk., 105:76.

Prineas, J. W., 1975, Pathology of the early lesions in multiple sclerosis, Hum.Pathol., 6:531.

Prineas, J. W., and Connell, F., 1978, The fine structure of chronically active multiple sclerosis plaques, Neurology, 28:68.

Prinease, J. W., and Connell, F., 1979, Remyelination in multiple sclerosis, Ann.Neurol., 5:22.

Prinease, J. W., and Raine, C. S., 1976, Electron microscopy and immuneperoxidase studies in early multiple sclerosis lesions, Neurology, 26:29.

Prinease, J. W., and Wright, R. G., 1978, Macrophages, lymphocates and plasma cells in the perivascular compartment in chronic multiple sclerosis, Lab.Invest., 38:409.

Putnam, T. J., 1935, Studies in multiple sclerosis. IV. Encephalitis and sclerotic plaques produced by venular obstruction, Arch. Neurol., 33:929.

Raine, C. S., and Bornstein, M. B., 1970, Experimental allergic encephalomyelitis: and ultrastructural study of experimental demyelination in vitor, J.Neuropathol.Exp.Neurol., 29:177.

Raine, C. S., Traugott, U., and Stone, S. H., 1978, Glial bridges and Schwann cell migration during chronic demyelination in the CNS, J.Neurocytol., 7:541.

Raine, C. S. Scheinberg, L., and Waltz, J. M., 1981, Multiple sclerosis: oligodendroglia survival and proliferation, Lab.Invest., 45:534.

Schwerer, B., Egghart, M., and Kitz, K., 1983, Serum antibodies against central nervous system (CNS) antigens in chronic relapsing experimental allergic encephalomyelitis (crEAE), in: "Protides of the Biological Fluids," H. Peeters, ed., Pergamon Press, Oxford, p. 1983.

Seil, F. J., Rauch, H. C., and Einstein, E. R., 1973, Myelination inhibition factor: its absence in sera from subhuman primates sensitized with myelin basic protein, J.Immunol., 111:96.

Seitelberger, F., 1960, Histochemistry of demyelinating diseases proper including allergic encephalomyelitis and Pelizäus-Merzbacher's disease, in: "Modern Scientific Aspects of Neurology," J. N. Cummings, ed., Arnold, London.

Seitelberger, F., 1973, Pathology of multiple sclerosis, Ann.Clin.Res., 5:337.

Seitelberger, F., 1982, Comparative neuropathology of oriental and western multiple sclerosis cases, in: "Multiple Sclerosis East and West," Y. Kuroiwa, and L. T. Kurland, eds., Kyushu University Press, Fukuoka.

Siemerling, E., and Raecke, E., 1914, Beitrag zur Klinik und Pathologie der multiplen Sklerose mit besonderer Berücksichtigung ihrer Pathogenese, Arch.Psychiat.Nervenkr., 53:385.

Steiner, G., 1931, "Krankheitserreger und Gewebsbefund bei multipler Sklerose," Springer, Berlin.

Stone, S. H., and Lerner, E. M., 1965, Chronic disseminated allergic encephalomyelitis in guinea pigs, Ann.N.Y.Acad.Sci., 122:227.

Suzuki, K., Andrews, J. M., and Waltz, J. M., 1969, Ultrastructural studies of multiple sclerosis, Lab.Invest., 20:444.

Traugott, U., Shevach, E., and Chiba, J., 1982, Chronic relapsing experimental allergic encephalomyelitis: Identification and dynamics of T and B cells within the central nervous system, Cell Immunol., 68:261.

Traugott, U, Reinherz, E., and Raine, C. S., 1983, Multiple sclerosis: Distribution of T-cell subsets within active chronic lesions, Science, 219:308.

Wisniewski, H. M., and Keith, A. B., 1977, CHronic relapsing experimental allergic encephalomyelitis - and experimental model of multiple sclerosis, Ann.Neurol., 1:144.

Wisniewski, H. M., and Lassmann, H., 1983, Etiology and pathogeneis of monophasic and relapsing inflammatory demyelination - human and experimental, Acta.Neuropathol., Suppl.9:21.

Wisniewski, H. M., Lassmann, H., and Brosnan, C. F., 1982, Multiple sclerosis: immunological and experimental aspects, in: "Recent Advances in Clinical Neurology," W. B. Mathews, and G. H. Glaser, eds, Churchill-Livingstone, London, vol.3, p.95.

WHAT EXPERIMENTAL ALLERGIC ENCEPHALOMYELITIS TEACHES US ABOUT MULTIPLE SCLEROSIS

Ellsworth C. Alvord, Jr.

Laboratory of Neuropathology, Department of Pathology
University of Washington School of Medicine
Seattle, Washington 98195

After 37 years of study of experimental allergic encephalomyelitis (EAE) and multiple sclerosis (MS) I obviously know more about these conditions than I did when I began, but two questions arise: 1) Could I have predicted where I am now sufficiently clearly to have convinced any granting agency to support my research for these 37 years, and 2) from my vantage point can I predict how and when the puzzle will be solved? Of course, the answer to both questions is no - the first, emphatically no; the second, more questioningly so. I have the impression I am looking over a photographer's shoulder, watching a picture being developed, with the more heavily exposed portions appearing first, apparently randomly scattered and only slowly coalescing to provide some details while the whole picture is still not recognizable.

In such a situation I can only tell you where I see us now, where I think we can progress rapidly and possibly where the big holes remain.

The year 1983 is too important an anniversary to allow it to pass without specific comment: the centennial of Pasteur's invention of rabies vaccine, the semi-centennial of River's proof that the "neuroparalytic accidents" of this rabies vaccine were due to the neural contaminants, and the quarter-centennial of Kies's extraction of myelin basic protein (BP) and proof that this was the specifically responsible neural antigen.

There are those purists who would wait another couple of years for the true centennial, but the pace of Pasteur's laboratory discoveries 100 years ago is difficult to appreciate today. In the 5 years centered around 1883 Louis Pasteur and his collaborators

(Chamberland, Roux and Thuillier) succeeded not only in cultivating but also in subduing the virus of rabies, in a wide variety of experimental animals as well as in humans, thereby setting the stage for the next century of research on what we now know as EAE.

The first 50 years of studies of the "neuroparalytic accidents" of rabies vaccine were the slowest since only a few investigators, such as Koritschoner and Schweinburg (1925) and Miyagawa and Ishii (1926), came up with what we now know to be essentially the right answer. It remained for Rivers and his collaborators (Schwentker, Sprunt and Berry) to show not only that the complication was due to a neural contaminant but also that demyelination was a prominent feature. That they mentioned Schilder's disease, not MS, is one of those ironies of fate.

The next 10 years must have been the most frustrating, since the concept that many demyelinating diseases might be due to an allergy comparable to that inducing EAE could hardly be further refined with such a tedious experimental design requiring so many injections over such a long period of time. Not until Jules Freund developed his adjuvants 40 years ago could the experimental mode be manipulated.

If the pace of the first 60 years was slow and frustrating, the pace of the last 40 years has certainly been accelerating and the results increasingly productive. In the decade between the meeting of the Association for Research in Nervous and mental Diseases on MS and our first symposium on EAE, that is, the decade including most of the 1950's, modest advances occurred, especially in identifying BP, but the really significant advances have occurred in the past 25 years, as illustrated in the graph below, recording my impressions of the ups and downs of the possible relationships between EAE and MS. As you can see, I am today about 80% convinced that EAE is a most appropriate model of MS.

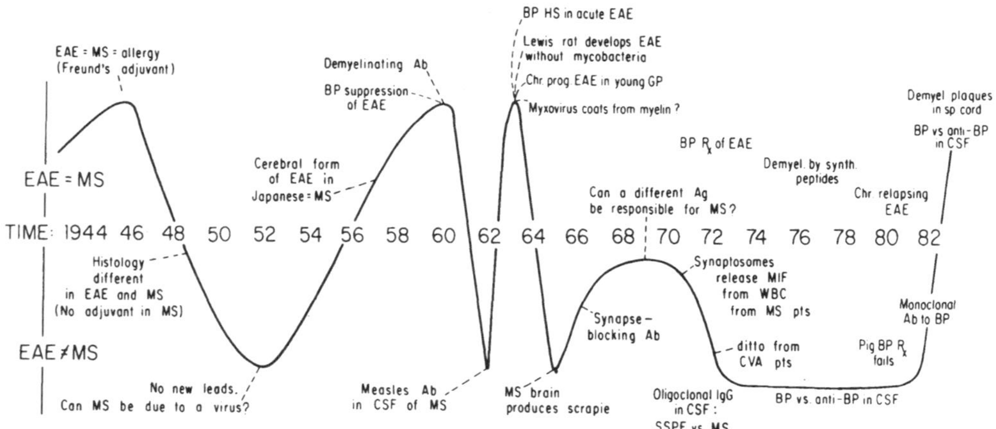

Fig. 1.

At our recent symposium on "How Good a Model of MS is EAE Today?" in Seattle last July some 65 papers were presented and will soon be published by Alan R. Liss, Inc., N. Y. and I will draw on some of these papers-in-press to round out the story.

The 4 major areas of our own current research are:

1 the definition of the onset of EAE by clinical, immunochemical and electrophysiological changes (Alvord et al., 1983a)
2 a comparison of cerebrospinal fluid changes in EAE and MS (Alvord et al., 1983b)
3 a morphologic comparison of 3 models of chronic remitting-relapsing EAE with MS (Shaw and Alvord, 1983)
4 an analysis of antigenic determinants in myelin basic protein by suing monoclonal antibodies (Hruby et al., 1983).

Time does not permit me to summarize each of these. Instead, since Dr. Wisniewski and Dr. Seitelberger have already presented their views on the relationships between EAE and MS, perhaps the best approach for me to take will be to outline what I have seen, paying special attention to where I disagree with each of the previous speakers. I will make 7 points:

First of all, I would emphasize that ordinary EAE looks different in each species that I am familiar with: guinea pig, rat, rabbit, monkey and man. That is not to say that there are not many similarities which allow us to produce and diagnose EAE, but there are significant differences which make comparisons with MS difficult (Table 1).

So, in Table 1 are some minor differences which I would call to the attention of Drs. Wisniewski and Seitelberger. These differences may be only quantitative or technical, but I cannot speak of EAE without specifying the animal and the protocol.

Second, I would emphasize that chronic-relapsing EAE has been produced in each of 3 species (guinea pig, rat and monkey) with quite different protocols. It is not yet fair to speak of only one mechanism producing demyelination. Let me describe each of these:

One of the advantages of being a pathologist is the ability to compare, at least at one point in time, the morphologic appearances of diseases in humans and experimental animals. During the past decade it has been my privilege to be able to study dozens of cases of multiple sclerosis (MS) and experimental allergic encephalomyelitis (EAE) in hundreds of monkeys, both in its untreated and usually acute form (Shaw and Alvord, 1976) and in its treated and healed forms (Alvord et al., 1979a), as well as in its suboptimally treated and usually chronic remitting-relapsing form (Alvord et al., 1983a,b). In addition, two new forms of chronic remitting-relapsing

Table 1. A Comparison of Some Histologic Features of EAE in Different Species with MS.

	EAE			
	guinea pig	rat	monkey	MS
1. Acute: perivascular lymphocytes	++	++	++	+
perivascular demyelination	0-±	0	±	++
2. Hyperacute: hemorrhage, necrosis	0	++*	++*	++
3. Chronic: demyelination	++	++	++	+++
sudanophilic lipids	0		++	+++
swollen axons	+++	++	0±	0-±
gliosis	0		+	+++
necrosis	0		+	±

* with pertussis
** with or without pertussis

EAE have been produced in guinea pigs and rats (Driscoll et al., 1982). These 3 experimental models suggest that pure demyelination can be produced, most likely as an intermediate degree of damage between forces that can produce necrosis at one extreme and non-destructive inflammation at the other.

The monkey model, as described by Alvord et al., (1983a,b), involves Macaca fascicularis monkeys sensitized to homologous (monkey) myelin basic protein (BP) in Freund's complete adjuvants (CFA) and treated suboptimally with BP and corticosteroids in saline in osmotic minipumps implanted subcutaneously.

The guinea pig model, as described by Driscoll et al., (1982), uses adult strain 13 guinea pigs actively challenged with homologous (guinea pig) spinal cord in CFA 1 day after receiving homologous BP-sensitized strain 13 lymph node cells which had been cultured for 24 hours in the presence of BP and an additional nonspecific stimulus (a mixed lymphocyte reaction) provided by normal strain 2 peritoneal exudate cells. These cells were given in suboptimal numbers; i.e., at 10-20% of the numbers required for passive transfer of acutely fatal EAE (Driscoll et al., 1979; 1982). Half of the animals were completely protected but half developed chronic varieties of EAE: progressive-fatal, remitting-relapsing or chronic-stable. The animals were sacrificed at intervals from 1 to 6 months.

The rat model of Namikawa (unpublished) follows the technique of Richert et al., (1981a,b) and uses Lewis rats actively challenged with homologous (rat) spinal cord in CFA. THis induces acute EAE followed by recovery. At 30 days the rats received passive transfer of guinea pig BP-sensitized rat lymph node and spleen cells, which induced a second attack of EAE followed by recovery. Similar passive transfers were given at 50 and 70 days, but each attack of EAE was followed by slower and less complete recovery. The animals were sacrificed at 90 days.

In general, each animal was sacrificed under nembutal anesthesia and perfused with fixative (10% formalin with neutral buffer). Representative blocks of brain and spinal cord were stained for myelin with gallocyanin-Darrow red or luxol fast blue, periodic acid-Schiff and hematoxylin and other stains as indicated.

Ordinary untreated EAE in the monkey is very frequently hyperacute, with hemorrhagic necrosis and extensive polymorphonuclear leucocytic exudate in the brain more than the spinal cord (Shaw and Alvord, 1976). Treatment with BP and an auxiliary factor (penicillin for M. mulatta and steroid for M. fascicularis) for a week or two can be completely successful (Shaw et al., 1976; Alvord et al., 1979), but cessation of treatment as soon as recovery has occurred (usually within a few days) allows another attack of EAE to occur, so that most of our monkeys have at least 2 and some 3 or even 4 attacks of EAE over a course of 2 to 3 months. The lesions in these chronic remitting-relapsing monkeys are very frequently demyelinating, the edges rather sharply defined, and only the center of the lesions shows evidence of old necrosis with loss of axons. The demyelinated axons are not noticeably swollen. Sudanophilic myelin breakdown products are prominent in frozen sections stained with oil-red-O. Plump reactive astrocytes are also prominent within the lesion.

Ordinary EAE in guinea pigs is typically predominantly inflammatory with perivenous lymphocytes in both brain and spinal cord. Demyelination is scanty but definite in about 15% of unselected animals, especially in animals showing disease of short incubation period and duration (Alvord et al., 1975). In the chronic guinea pig, however, there is relatively little inflammation and the demyelination is extensive with long subpial plaques scattered throughout the length of the spinal cord. Macrophages loaded with myelin debris are prominent, throughout the lesions during the first 1 to 2 months and at the edges of the lesion as it heals during the 3rd and 4th months. No sudanophilia has been seen in frozen sections stained with oil-red-O. The axons are swollen but remain intact. Reactive astrocytes, if present, do not stain with Holzer's stain; indeed, the extreme swelling of the axons suggests that there is little if any space to be accounted for by astrocytic reaction. Swollen myelin sheaths and balls of myelin are prominent at the edges, within otherwise normal-appearing white matter, without obvious cellular infiltrate, in all states except the completely healed. This pattern suggests a radial expansion of the demyelinating process.

Ordinary EAE in Lewis rats is also typically inflammatory, with perivenous lymphocytes predominantly in the spinal cord, especially caudally in the gray matter. In the present animals with repeated attacks induced by repeated passive transfer of BP-sensitized cells, however, the lesions closely resemble those in the chronic guinea pigs, with long subpial patches of extensive demyelination. Macrophages contain abundant myelin debris. The axons are intact, only slightly separated by the macrophages, swollen only a little if at all. Reactive astrocytes are not prominent.

Each of the 3 experimental models has its own characteristic features which make each recognizably different. The common features include repeated attacks of EAE (in the monkey and guinea pig models deliberately induced by suboptimal treatments or numbers of cells) and striking demyelination (remarkably pure in the guinea pig and rat, still with central necrosis in the monkey). Two of the models (guinea pig and rat) use whole spinal cord in CFA as an active challenge in addition to BP-sensitized cells, but one (monkey) uses BP only. Thus, although other antigens may be important, BP can suffice. Here is a major point where I disagree with Dr. Wisniewski: how many antigens are necessary to produce chronic relapsing EAE?

More specifically, is BP the whole antigen for EAE? We know that it can induce ordinary EAE and protect against ordinary EAE even when EAE is induced with whole CNS (Alvord et al., 1965). We know that suboptimal treatments of monkeys that have received no antigen other than BP can induce relapsing EAE with relatively prominent demyelination (Shaw and Alvord, 1983). The demyelination is not so pure as we see it in the guinea pig or rat, where whole CNS is also used, and where for the moment I have to admit the real possibility of another antigen. So, on this point I can as yet only partly disagree with Dr. Wisniewski, but I would point out that BP-sensitive cells capable of transferring ordinary EAE persist in such chronic guinea pigs for months (Driscoll et al., 1982), as though the particulate BP-complex in whole CNS persists at the site of innoculation longer than BP does in solution (Tabira et al., 1983). There may be, therefore, again merely quantitative, rather than qualitative, factors. As for galactocerebroside being the antigen inducing demyelinating antibody, Seil and Agrawal (1983) have pointed out that it evokes antibodies that cause peripheral as well as central myelin to lyse and that a chloroform-methanol insoluble "protein" may be responsible for the specific CNS demyelinating antibody.

For my third point, I would like to return to the pathogenesis of EAE: How does the sensitized lymphocyte recognize the target? BP is now generally considered to be on the inside of the oligodendroglial membrane, thus protected by the outer part of this cell's membrane, to say nothing of the glial foot plates, basement membrane and endothelial cells that separate the CNS from the circulating leucocytes.

Dr. Wisniewski would have us believe that local leakage of BP from myelin (as part of the metabolic cycle that allows BP to escape into the CSF) occurs through venules into the blood and that the spotty presence of BP on the endothelial surface of these venules attracts the sensitized lymphocytes to that particular focus. If there is BP in the blood - and I remain unconvinced that there is - why would it not desensitize the lymphocytes? No, I'd prefer to see the lymphocytes randomly migrating into and through cells, liberating lymphokines whenever they see the antigen to which they are sensitized, the lymphokines recruiting 90% of the cells in the inflammatory exudate.

To summarize at this point, let me say that I recognize a spectrum of clinical and histological reaction in these different species, which can be understood by oversimplifying and identifying them as hyperacute, acute and chronic. As I proposed 20 years ago, the simplest hypothesis that I see is that these represent changes in the balance of power between two major forces, hypersensitivity and immunity in the older terminology, helper and suppressor cells in the newer terminology. The major question, however, is whether these two forces represent qualitative or merely quantitative differences, as Knight (1982) has proposed. At the moment I am biased by the observations of Driscoll et al., (1982), who could convert passive transfer of EAE to passive transfer of protection against EAE merely by transferring fewer cells.

Similarly, I see demyelination as part of a quantitatively defined spectrum of damage intermediate between necrosis and simple inflammation, much as Hurst (1944) maintained many years ago.

This general conclusion then leads me to my fourth point and forces me to re-evaluate the definition of MS. In retrospect, the classical definition of MS as consisting of relatively large areas of relatively pure demyelination occurring as multiple attacks over time and space must represent a wastebasket, catching all the possible variations on the theme, and including a spectrum from acute disseminated encephalomyelitis and acute hemorrhagic-necrotic encephalomyelitis, including many cases of Balo's (Courville, 1970) and Devic's syndromes (Cloys and Netsky, 1970), through acute MS to relapsing MS and chronic MS. In the human species, in contrast to the other species studied experimentally, the most common reaction in most parts of the world is what we recognize statistically as "MS"; but in certain parts of the world, such as Japan and South America, the most common reaction is Devic's syndrome (Shiraki, 1968).

In my opinion, the most important question we have to face up to is whether MS is a disease per se or a syndrome. And as I have indicated above, I believe we are dealing with a syndrome, in which the several equivalents of the experimental protocols in the various species are operating upon the genetically non-homogeneous human populations to produce one or another portion of the clinical-histological spectrum.

For my fifth point and derived from this background, let me continue with some other observations which may have important implications for further investigations and interpretations, especially with regard to the question whether BP is the right antigen for MS.

Our studies of monoclonal antibodies to BP (Hruby et al., 1983) have reinforced the observations of many investigators that BP has many antigenic as well as many encephalitogenic determinants. In general, the encephalitogenic sites are species-specific, each recipient developing EAE following sensitization to a different sequence of amino-acid residues (Alvord, 1983). Although sequences as short as 6 residues may be encephalitogenic (Nagai, 1978), longer sequences are more effective, and perhaps as many as 20 may be necessary to mimic the whole BP when compared on a molar basis (Alvord, 1983). Some of these determinants are located in relatively invariant regions of the molecule, others in hypervariable regions, so that I see no rule which would permit a prediction of which region(s) may be encephalitogenic for humans. Indeed, based on the different susceptibilities of inbred strains of mice (Chou et al., 1983) and guinea pigs (Kies and Driscoll, 1982), I suspect that the genetically non-homogeneous humans may react to several different determinants.

The encephalitogenic sites so far defined can be summarized in the following diagram:

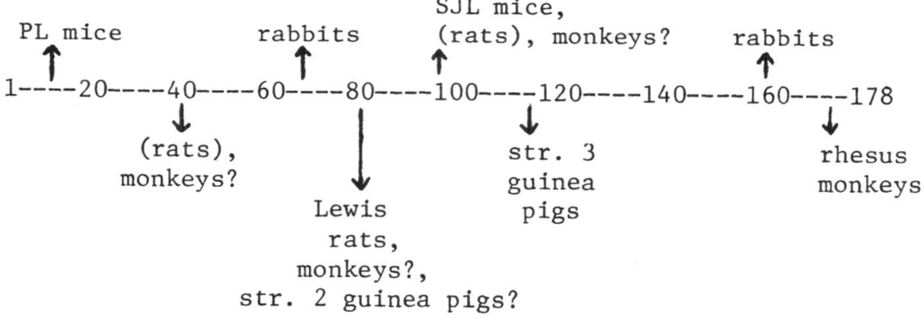

The parentheses indicate minor determinants, the ?'s indicate the lack of precise localization.

The other determinants, i.e., antibody-binding sites, are at least as interesting in their own right. Those so far identified by monoclonal antibodies are listed in Table 2.

As an aside, I should point out an interesting corollary of these differences, namely, that some antibody-binding determinants are more affected by fixation than others, and that some myelinated fibers appear to require better fixation than is provided by simple 10% formalin (4% formaldehyde, neutral buffered). For example, the deep horizontal fibers of the molecular layer of the cerebellum do not stain with monoclonal antibody-immunoperoxidase or even chemically unrelated common myelin sheath stains (such as luxol fast blue or gallocyanin) unless the specimen is fixed in formaldehyde containing metallic ions (copper or mercury), as in Susa ($HgCl_2$) or Bouin-Hollande-sublimate (Cu acetate). I do not yet know the implications of this observation for MS, but I suspect 1) that abnormal myelin will be subject to similar requirements for a fixative which is better than the classical formaldehyde and 2) that better stains may modify our diagnostic opinions!

Sixth, is BP the right antigen for MS? We know that pig BP failed as a treatment of MS (Romine and Salk, 1983); but, by analogy with the species-restrictions of some of our monoclonal antibodies and with the lack of encephalitogenicity of bovine BP for Lewis rats, I suspect that the human encephalitogenic determinant(s) may not be present in pig BP (i.e., that the homologous sequence is not the same in pig as in human BP). We know that BP in the CSF is not specific for MS and that anti-BP antibodies in the CSF are not specific for MS (Alvord et al., 1983b), but we wonder if we are testing for the right antigenic determinant. We know that the demyelinating factor in human sera is not specific for MS, that it might not even be an antibody, and that it may represent an epiphenomenon not at all related to the demyelinating antibodies of whole-CNS-induced EAE (Bradbury and Suckling, 1983).

Table 2. Monoclonal Antibody-Reactive Sites in BP.

Residues*	Class of MAb	Degree of Species-Restriction
1-14	rat IgG	0?
90-99	rat IgG	+
86-100	mouse IgM	0?
114-124	rat IgG	++
131-140	mouse IgG, IgM	++++

* numbering based on a hypothetical molecule with 178 positions including Phe-Phe @ 44-45 and 91-92; Trp @ 118; rat-14kDa deletion of 120-161 (Martenson, 1983).

So, we have some powerful arguments against BP being the right antigen for MS.

Seventh, and last, do we have as powerful arguments against EAE as the best model for MS? Here the evidence is more suggestive that we are increasingly gaining experimental control of the mechanisms producing demyelination. It is of interest to compare these lesions with those in humans with MS, where several forms have been described: 1) phagocytosis of normal-appearing myelin (Alvord, 1977; Prineas and Connell, 1978), 2) phagocytosis of abnormal appearing myelin (Zimmerman and Netsky, 1950), 3) confluence of small perivascular lesions (Prineas, unpublished) and 4) large "pre-plaques" identified by accumulations of IgG-laden macrophages (Prineas, unpublished). The third is the rarest but bears the closest resemblance to ordinary EAE in practically all animals. The first is the most debated, since it implies a primary attack by macrophages, stripping off otherwise normal myelin. The second is also debated, since it implies a primarily noncellular (? viral, ? antibody, ? enzyme) damage to myelin; it is quite easily seen in the present guinea pig model but not in the rat or monkey. Of the 4 patterns that have been described in MS, we have seen 2 in EAE: 1) a confluence of small perivascular lesions and 2) a radial expansion with phagocytosis of abnormal myelin. The radial expansion with phagocytosis of normal myelin and the large "pre-plaque" accumulation of IgG-laden macrophages have not yet been reproduced. .

Although a review of my first publication (Alvord and Stevenson, 1950) makes it clear that some of the chronic lesions were accepted then as parts of the spectrum of ordinary EAE in guinea pigs, it is now obvious that different aspects of these lesions are coming under more strict experimental control. In particular, remarkably pure demyelination can be produced.

Thus, I return to the concept that MS is a syndrome, with multiple demyelinating mechanisms, some of which can be mimicked by EAE. But the most frustrating point is that we need more information about the early lesions of MS: does it involve leucocytes or not? We could eliminate a number of hypotheses if we knew the answer to this simple question. In EAE we are reasonably confident that leucocytes are primarily involved, but in MS we are not so sure. We can occasionally see lesions in frozen sections stained with oil-red-O and hematoxylin (Alvord, 1977), lesions which by analogy with strokes can be only several hours to a few days old and are still evolving, but is every active lesion of this type, with macrophages phagocytosing normal-appearing myelin? Since other investigators describe other lesions as being active, I conclude that we do not know the whole story.

In conclusion, then while we know the etiology, pathogenesis, prevention and treatment of EAE, we do not know any of these aspects of MS. Even so, I am still 80% convinced that EAE is the best model we have.

Acknowledgment

Supported in part by research grant RG-805-E-20 from the National Multiple Sclerosis Society and grant 3P51-RR-00166-1 from the National Institutes of Health to the Regional Primate Center.

REFERENCES

Alvord, E. C., Jr., 1965, Pathogenesis of experimental allergic encephalomyelitis: Introductory remarks, Ann.N.Y. Acad.Sci., 122:245-255.

Alvord, E. C., Jr., 1977, Demyelination in experimental allergic encephalomyelitis and multiple sclerosis, in: "Slow Virus Infections of the Central Nervous System, Investigational Approaches to Etiology and Pathogenesis of These Diseases," V. ter Meulen, and M. Katz, eds., Springer-Verlag, New York, pp. 166-185.

Alvord, E. C., Jr., 1983, Species-restricted encephalitogenic determinants, in: "Is EAE a Good Model for MS?", E. C. Alvord, Jr., M. W. Kies, and A. J. Suckling, eds., Alan R. Liss, New York.

Alvord, E. C., Jr., and Stevenson, L. D., 1950, Experimental production of encephalomyelitis in guinea pigs, Res.Publ.Assoc. Res.Nerv.Ment.Dis., 28:99-112.

Alvord, E. C., Jr., Shaw, C. M. Hruby, S., and Kies, M. W., 1965, Encephalitogen-induced encephalomyelitis: Prevention, suppression and therapy, Ann.N.Y. Acad.Sci., 122:333-345.

Alvord, E. C., Jr., Shaw, C. M., Hruby, S., Peterson, R., and Harvey, F. H., 1975, Correlation of delayed skin hypersensitivity and experimental allergic encephalomyelitis induced by synthetic peptides, in: "The Nervous System", D. B. Tower, ed., Raven Press, New York, pp. 647-653.

Alvord, E. C., Jr., Shaw, C. M., and Hruby, S., 1979, Myelin basic protein treatment of experimental allergic encephalomyelitis in monkeys, Ann.Neurol., 6:469-473.

Alvord, E. C., Jr., Shaw, C. M., Hruby, S., Sires, L. R., and Slimp, J. C., 1983a, The onset of experimental allergic encephalomyelitis as defined by clinical, electro-physiological and immunochemical changes, in: "Is EAE a Good Model for MS?", E. C. Alvord, Jr., M. W. Kies, and A. J. Suckling, eds., Alan R. Liss, New York.

Alvord, E. C., Jr., Hruby, S., Shaw, C. M., and Slimp, J., 1983b, Myelin basic protein and its antibodies in the cerebrospinal fluid in experimental allergic encephalomyelitis, multiple sclerosis and other diseases, in: "Is EAE a Good Model for MS?",

E. C. Alvord, Jr., M. W. Kies and A. J. Suckling eds., Alan R. Liss, New York.

Bradbury, K., and Suckling, A. J., 1983, The nature of factors in chronic relapsing EAE and MS sera which induce myelino-toxicity and cellular changes in organ culture, in: "Is EAE a Good Model for MS?", E. C. Alvord, Jr., M. W. Kies, and A. J. Suckling, eds., Alan R. Liss, New York.

Chou, C.-H. J., Shapira, R., and Fritz, R. B., 1983, Further delineation of encephalitogenic determinant for PL/J and (SJLxPL)Fl mice, in: "Is EAE a Good Model for MS?", E. C. Alvord, Jr., M. W. Kies, and A. J. Suckling, eds., Alan R. Liss, New York.

Cloys, D. E., and Netsky, M. G., 1970, Neuromyelitis optics, in: "Handbook of Clinical Neurology," P. J. Vinken, and G. W. Bruyn, eds., North-Holland Publishing Company, Amsterdam, 9:426-436.

Courville, C. B., 1970, Concentric sclerosis, in: "Handbook of Clinical Neurology," P. J. Vinken, and G. W. Bruyn, eds., North Holland Publishing Company, Amsterdam, 9:437-451.

Driscoll, B. F., Kies, M. W., and Alvord, E. C., Jr., 1979, Enhanced transfer of experimental allergic encephalomyelitis with strain 13 guinea pig lymph node cells: Requirement for culture with specific antigen and allogeneic peritoneal exudate cells. J.Immunol., 125:1817-1822.

Driscoll, B. F., Kies, M. W., and Alvord, E. C., Jr., 1982, Suppression of acute experimental allergic encephalomyelitis in guinea pigs by prior transfer of suboptimal numbers of EAE-effector cells: Induction of chronic EAE in whole tissue-sensitized guinea pigs, J.Immunol., 128:635-638.

Hruby, S., Alvord, E. C., Jr., Martenson, R. E., Deibler, G. E., Hickey, W. F., and Gonatas, N. K., 1983, Epitopes in myelin basic protein reactive with monoclonal antibodies, in: "Is EAE a Good Model for MS?", E. C. Alvord, Jr., M. W. Kies, and A. J. Suckling, eds., Alan R. Liss, New York.

Hurst, E. W., 1944, A review of some recent observations on demyelination, Brain, 67:103-124.

Kies, M. W., and Alvord, E. C., Jr., eds., 1959, "Allergic' Encephalomyelitis," Charles C. Thomas, Springfield.

Kies, M. W., and Driscoll, B. F., 1983, Immunologic reactivity of myelin basic protein (BP) in inbred guinea pigs. Protides of the Biological Fluids, 30:115-118.

Knight, S. C., 1982, Control of lymphocyte stimulation in vitro: 'Help' and 'suppression' in the light of lymphoid population dynamics, J.Immunol.Meth., 50:R51-R63.

Koritschoner, R., and Schweinburg, F., 1925, Klinische und experimentelle Boebachtugen über Lähmungen nach Wutschutzimpfung, Zts f. Immunitätsf, 42:217-282.

Martenson, R. E., 1983, Myelin basic protein speciation, in: "Is EAE a Good Model for MS?", E. C. Alvord, Jr., M. W. Kies, and A. J. Suckling, eds., Alan R. Liss, New York.

Miyagawa Y, and Ishii, S., 1926, On the influence of the constituents of central nerve cells parenterally injected on the living organism, Jap.J.Exp.Med. (Sci. Reports Govt. Inst. Inf. Dis., Tokyo Imp. Univ.), 5:331-370.

Nagai, Y., Akiyama, K., Suzuki, Kusumoto, S., Ikuta, F., and Takeda, S., 1978, Minimum structural requirements for encephalitogen and for adjuvant in the induction of experimental allergic encephalomyelitis, Cell.Immunol., 35:158-167.

Pasteur, L., 1885, Méthode pour prévenir la rage aprés morsure, Comptes rendus des séances de l'acadèmie des sciences, 101:765-774.

Pasteur, L., Chamberland, Roux, Thuillier, 1881, Sur la rage, Comptes rendus des séances de l'académie des sciences, 92:1259-1260.

Prineas, J. W., and Connell, F., 1978, The fine structure of chronically active multiple sclerosis plaques, Neurology, 28(Suppl):68-75.

Richert, J. R., Kies, M. W., Alvord, E. C., Jr., 1981a, Enhanced transfer of experimental allergic encephalomyelitis with Lewis rat lymph node cells, J.Neuroimmunol., 1:195-203.

Richert, J. R., Driscoll, B. F., Kies, M. W., and ALvord, E. C., Jr., 1981b, Experimental allergic encephalomyelitis: Activation of myelin basic protein-sensitized spleen cells by specific antigen in culture, Cell.Immunol., 59:42-53.

Rivers, T. M., and Schwentker, F. F., 1935, Encephalomyelitis accompanied by myelin destruction experimentally produced in monkeys, J.Exp.Med., 61:689-701.

Rivers, T. M., and Schwentker, F. F., 1935, Encephalomyelitis accompanied by myelin destruction experimentally produced in monkeys, J.Exp.Med., 61:689-701.

Romine, J. S., and Salk, J., 1983, A study of myelin basic protein as a therapeutic probe in patients with multiple sclerosis, in: "Multiple Sclerosis," J. F. Hallpike, C. W. M. Adams, and W. W. Tourtellotte, eds., University Press, Cambridge, pp.621-630.

Seil, F. J., and Agrawal, H. C., 1983, Serum antimyelin factors in experimental allergic encephalomyelitis and multiple sclerosis, in: "Is EAE a Good Model for MS?", E. C. Alvord, Jr., M. W. Kies, and A. J. Suckling, eds., Alan R. Liss, New York.

Shaw, C. M., and Alvord, E. C., Jr., 1976, Treatment of experimental allergic encephalomyelitis in monkeys, II. Histopathological studies, in: "The Aetiology and Pathogenesis of the Demyelinating Diseases," H. Shiraki, T. Yonezawa, and Y. Kuroiwa, eds., Japan Science Press, Tokyo, pp. 377-395.

Shaw, C. M., and ALvord, E. C., Jr., 1983, A morphologic comparison of three experimental models of experimental allergic encephalomyelitis with multiple sclerosis, in: "Is EAE a Good Model for MS?", E. C. Alvord, Jr., M. W. Kies, and A. J. Suckling, eds., Alan R. Liss, New York.

Shaw, C. M., Alvord, E. C., Jr., and Hruby, S., 1976, Treatment of experimental allergic encephalomyelitis in monkeys. I. Clinical studies, in: "The aetiology and Pathogenesis of the

Demyelinating Diseases," H. Shiraki, T. Yonezawa, and Y. Kuroiwa, Japan Science Press, Tokyo, pp. 367-376.

Shiraki, H., 1968, The comparative study of rabies postvaccinal encephalomyelitis and demyelinating encephalomyelitides of unknown origin, with special reference to the Japanese cases, in: "The Central Nervous System, Some Experimental Models of Neurological Diseases," O. T. Bailey, ed., The Williams & Wilkins Company, Baltimore, pp. 87-123.

Tabira, T., Itoyama, Y., and Kuroiwa, Y., 1983, The rôle of locally retained antigens in chronic relapsing experimental allergic encephalomyelitis in guinea pigs, in: "Is EAE a Good Model for MS?", E. C. Alvord, Jr., M. W. Kies, and A. J. Suckling, eds., Alan R. Liss, New York.

Zimmerman, H. M., and Netsky, M. G., 1950, The pathology of multiple sclerosis, Res.Publ.Assoc.Res.Nerv.Ment.Dis., 28:271-312.

ANTIGEN AND ANTIBODY STUDIES IN DEMYELINATING DISEASE

Raymond P. Roos[1] and David H. Mattson[2]

[1]University of Chicago Medical Center
Chicago, Illinois
[2]University of Pennsylvania Medical Center
Philadelphia, Pennsylvania

INTRODUCTION

A potential clue to our understanding of the pathogenesis of multiple sclerosis (MS) is the presence in MS cerebrospinal fluid (CSF) and brain eluates of heterogeneously distributed IgG, i.e. oligoclonal IgG bands. The first description of abnormal IgG bands in MS was by Lowenthal (1960) who observed that CSF IgG from MS had "restricted heterogeneity" when analyzed by agar gel electrophoresis (AGE). Laterre (1965) referred to these bands as "oligoclonal" in order to emphasize the derivation of these bands from several clones of antibody synthesizing cells. The monoclonality of the bands has been supported by evidence that individual bands possess a single light chain type (Link and Laurenzi, 1978; Mattson et al., 1982a) and single idiotype (Ebers et al., 1979; Nagelkerken et al., 1980; Gerhard et al., 1981; Tachovsky et al., 1982).

In order to elucidate the MS humoral IgG response, we have found it valuable to compare the characteristics of the CSF IgG pattern in MS with the pattern found in a persistent central nervous system (CNS) infection, subacute sclerosing panencephalitis (SSPE), and an autoimmune CNS disease, experimental allergic encephalomyelitis (EAE). Our investigations, as reported here, show that the MS humoral immune response is clearly different from that seen in SSPE patients and indicates a different fundamental pathogenesis of disease. Our studies suggest that perhaps much or all of the MS IgG is not related to a specific etiologic agent or autoimmunogen, i.e. it is "nonsense" antibody. Our unsuccessful search to date for a specific antigenic target for the MS IgG may result from the presence of nonsense antibody.

ANTIBODY STUDIES

IEF Technical Aspects

Our initial studies compared the techniques of AGE and isoelectric focusing (IEF). The separation of proteins by AGE is on the basis of net charge, size and shape, while IEF separation only depends on the isoelectric point (pI) of the protein. IEF has superior resolution compared to AGE, but it is more difficult to interpret. A main difficulty in interpretation stems from a background pattern of uniform markings seen even when normal serum or CSF is focused. These normal markings presumably result at least partly from "mini-steps" in a pH gradient which is not completely homogeneous and which reflects a finite number of ampholines. (Maxey and Palmer, 1975).

In order to increase our sensitivity and allow for immunochemical identification of the IgG, we originally used an immunoperoxidase overlay technique (Mattson et al., 1980a). More recently, we have employed an extremely sensitive silver staining method (Roos and Lichter, 1983) (Figure 1) which can be combined with an immunofixation step. The ability to resolve as little as 80 ng of IgG in specimens has enabled us to analyze unconcentrated CSF and brain eluates.

IEF Banding

Clinical Correlates. In our original studies (Mattson et al., 1981), we found that 88% and 100% of CSF from MS and SSPE patients had oligoclonal IgG bands respectively. In MS, but not SSPE, the number of bands detected by IEF correlated with increasing disease duration, CSF IgG and CNS IgG synthesis. Each patient's pattern was unique. Although the overall banding pattern remained fairly stable over time, some changes in band number and relative intensities did occur.

Banding Pattern. Densitometric scanning of IEF gels enabled us to make the following distinctions between MS, SSPE and normal CSF IgG IEF patterns: SSPE CSF oligoclonal bands were more peaked and prominent than MS bands; MS CSF had more "background" polyclonal IgG than normal CSF suggesting a local synthesis of polyclonal (as well as oligoclonal) IgG; SSPE CSF had less "background" polyclonal IgG than MS or even than normal CSF indicating that a large proportion of SSPE IgG was oligoclonal in distribution.

Light Chain Composition. We studied the light chain composition of MS and SSPE CSF IgG in order to verify the monoclonality of individual CSF bands and to contrast the findings in the two diseases (Mattson et al., 1982b). When studied by IEF followed by an immuno-

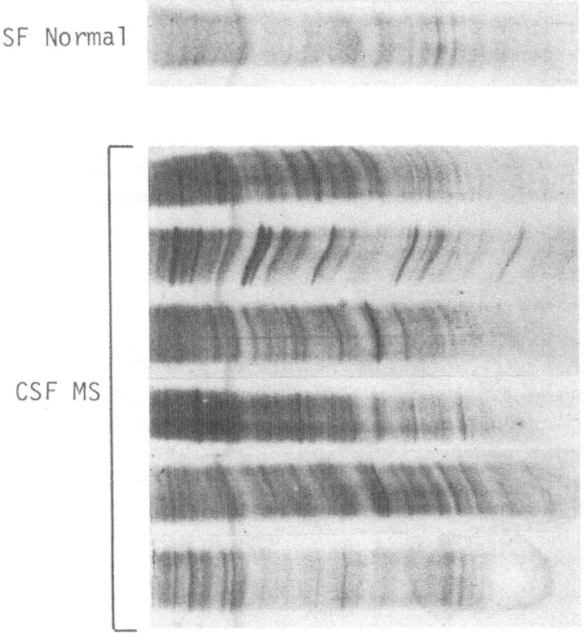

Fig. 1. IEF patterns in normal and MS CSF run on LKB Ampholine PAG plate gels, pH 3.5-9.5 and silver stained (Roos and Lichter, 1983). In this and all subsequent gels, only the cathodal end of the gel is shown. Bands corresponding to IgG are present in the MS specimens.

chemical overlay technique many of the bands in MS and SSPE CSF were found to contain only one light chain type, indicating their monoclonality. MS bands tended to be κ predominant, whereas SSPE bands had a more normal κ/λ ratio. The reason for the greater predominance of κ light chains in MS CSF is not certain, but may relate to the nature of the antigen or the age of the individual when the clone is generated. MS antigens may tend to generate κ predominant antibody responses, as reported in other antibody responses (Barundun et al., 1976). SSPE individuals usually have a primary measles infection early in life when κ/λ ratios are known to be lower than in adults (Barundun et al., 1976).

Occassional bands stained for λ light chain, but not IgG heavy chains; these bands presumably represent free light chains since IgM does not enter the polyacrylamide gels we used and since IgA focuses in another area of the gel. The presence of free λ light chains has been noticed by others in MS (Laurenzi et al., 1979) but their pathogenesis remains obscure, especially since there is an overall κ/λ predominance of the IgG bands. Occasional IgG bands stained for both κ and λ light chains, probably because of a limited pH gradient set up by the finite number of ampholines resulted in cofocusing of different bands.

In summary, our investigations demonstrated a general similarity between the light chain composition of SSPE and MS IgG bands, except for the more frequent predominance of IgG λ bands in MS than in SSPE.

Nonsense Antibody

In our investigations of MS and SSPE IgG bands, we compared the band patterns of serum, CSF and eluates from different areas of the brain from the same individual (Mattson et al., 1980b); Mattson et al., 1982a). In the case of MS, patterns from different regions and compartments from a single individual had some bands in common, but also different unique bands (Figures 2, 3); these results were recently reproduced by others (Mehta et al., 1982). In contrast, SSPE band patterns from different compartments of a single individual were extremely similar (Figure 4), suggesting a difference in the pathogenesis of IgG bands in the two diseases. In SSPE, there appears to be a common response of identical plasma cell clones to the same antigen, measles virus, in all the compartments. The difference in patterns seen in MS indicates that different plasma cell clones are present in these different areas; most likely these clones are responding to different antigens. The band that are unique to one compartment may represent nonsense antibody, i.e. IgG that is not relevant to disease pathogenesis; the bands in common in different regions may be directed against a relevant antigenic target.

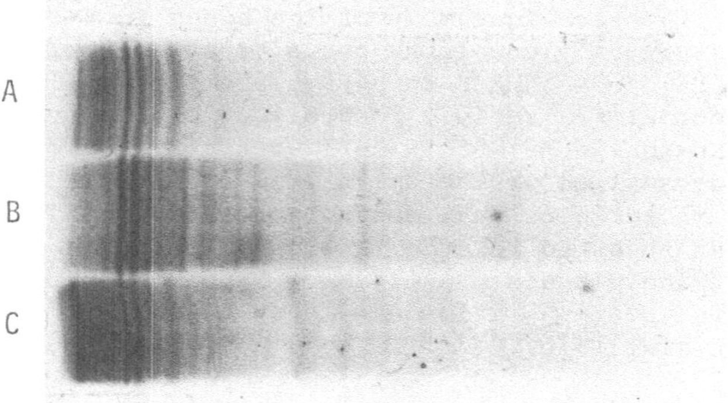

Fig. 2. IEF with immunoperoxidase anti-IgG overlay staining of brain eluates from MS case 1. A) acid eluate of pooled white matter; B) acid eluate of plaque 1; C) acid eluate of plaque 2. For details regarding the procedure see Mattson et al., 1980b, 1982a.

There are several other lines of data that we and others have presented which support the presence of nonsense antibody in MS. Our finding that MS CSF IEF band number correlates with disease duration (Mattson et al., 1981) suggests that, with time, antibody-synthesizing cell clones accummulate in the CNS, producing more bands. We know from studies with cultured mononuclear cells stimulated by pokeweed mitogen (Oger et al., 1981) that this non-specific stimulation of a limited number of B cell clones can produce an oligoclonal response. Some of the oligoclonal bands may blend together into the increased polyclonal IgG of MS CSF that we observed on IEF gels (Mattson et al., 1981). Norrby et al. (1974) reported that individual MS patients can have antibody to multiple viruses locally produced in the CNS. Activity of these varied antibodies can be localized to a small part of the oligoclonal CSF IgG bands as well as the polyclonal IgG with a sensitive electroimmunofixation method (Rostrom et al., 1981; Vartdal and Vandvik, 1982).

What determines which antibodies comprise the MS IgG response? Perhaps antigens which frequently invade the brain, such as measles virus, generate plasma cell clones which reside in the CNS for extended periods and occassionally become active as part of an antibody response directed against another target. In this connection, one might ask why SSPE does not display nonsense IgG. Evidence suggests that all of the SSPE IgG bands comprise antibodies that are only directed against measles virus: all the bands have activity against measles virus and can be absorbed out by measles virus, SSPE IgG banding patterns from different compartments are similar; IEF band

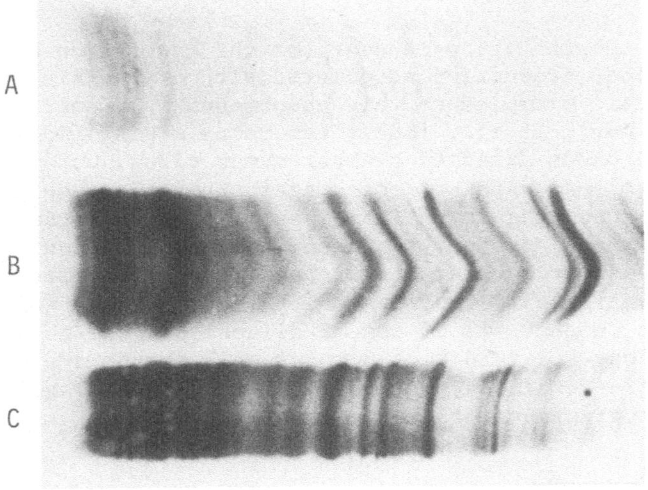

Fig. 3. IEF with immunoperoxidase anti-IgG overlay staining of MS specimens from case 2. A) acid eluates of pooled white matter; B) neutral saline eluate of pooled white matter; C) CSF.

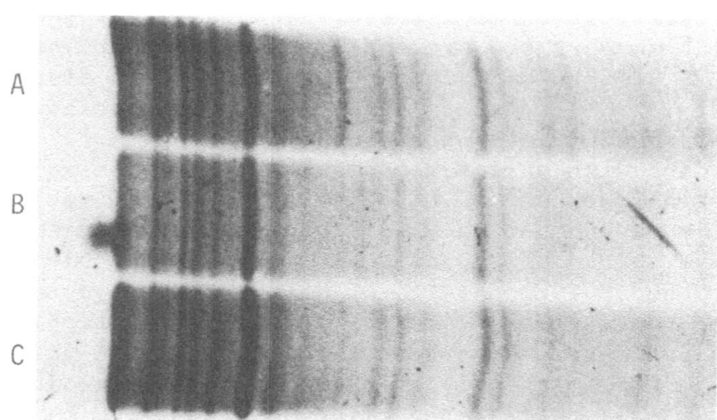

Fig. 4. IEF with immunoperoxidase anti-IgG overlay staining of brain eluates from SSPE case. A) acid eluate of region 1; B) acid eluate of region 2; C) acid eluate of region 3.

number does not correlate with increased CNS IgG synthesis in SSPE (as it does in MS) implying that with continuing hyperimmunization all SSPE oligoclonal IgG production goes into the same set of bands with anti-measles antibody. Perhaps the extreme intensity and chronicity of the measles virus antigenic stimulation leads to this "non-nonsense" antibody response in SSPE.

There is substantial precedent for the generation of a non-specific antibody production associated with a specific one. One of the most dramatic examples of this phenomenon was described in mumps meningitis (Vandvik et al., 1982); ten mumps meningitis patients had oligclonal IgG bands directed against mumps virus, but four had, in addition, oligclonal bands directed against measles, herpes and/or rubella virus appearing in variable association and order. In several experimental systems a nonsense antibody response has been found to antedate or even exceed a specific antibody response (see Mattson et al., 1982a for references). The important question that now remains is whether all of the MS oligoclonal IgG response is nonsense antibody - a result of B and T cell immune dysregulation - or whether there remains a portion which is directed against the relevant etiologic agent or autoimmunogen.

Oligoclonal IgG Bands in EAE

We initiated studies on EAE in order to determine whether bands were present in autoimmune diseases and additionally whether these bands might represent nonsense antibody. Our studies, performed in collaboration with Drs. Caroline Whitacre, Philip Paterson and Eugene

Day have involved acute EAE in the rabbit (Whitacre et al., 1981; Whitacre et al., 1982). Because hemoglobin bands focus near the pI or rabbit IgG, we chose to use AGE for our studies. Rabbits immunized with whole CNS tissue or myelin basic protein (MBP) displayed IgG bands in sera and CSF. Slices of the agar gels from electrophoresed serum and CSF of rabbits immunized with MBP were eluted and assayed for anti-MBP activity. There was no localization of MBP antibody activity to band-containing fractions. In addition, absorption with MBP did not change the banding pattern. In one experiment absorption with the BCG component of Freund's complete adjuvant also did not change the pattern.

Bands have also been described in sera and CSF from guinea pigs with chronic relapsing EAE (CREAE) (Glynn et al., 1982; Karcher et al, 1982). Absorption with M. tuberculosis removed most of the IgG bands. Although animals had evidence by immunoblotting of antibody against varied CNS polypeptides, including MBP, absorption with CNS homogenates did not change the banding pattern (Glynn et al., 1982). It may have been valuable to have specifically localized the M. tuberculosis antibody activity to the bands more precisely, for example, by elution and, in addition, to have ruled our non-specific absorption of IgG to the M. tuberculosis. My suspicion is that many of the bands in CREAE are, as reported, directed against M. tuberculosis, but that a smaller proportion of the bands are also directed against CNS antigens.

Although controversy exists as to the major antigenic target of bands found in EAE, the presence of M. tuberculosis antibody and the lack of correspondence between the bands and an autoimmunogenic neural antigen lend support to the concept that oligoclonal IgG may be irrelevant to the pathogenetic autoimmunogen, i.e. it may be nonsense antibody. As Glynn et al. (1982) hypothesize, "it may be that, by analogy, the oligoclonal IgG in an MS patient represents a summation of humoral immune responses to infectious agents, which in certain individuals, can act as immunological adjuvants.".

Antigen Studies

Although it is clear that some of the polyclonal and oligoclonal IgG is nonsense antibody, our working hypothesis has been that at least a portion of the oligoclonal "orphan" IgG (i.e., antibody looking for an antigenic target) has a relevant etiologic agent or autoimmunogen as its target.

Absorption Studies

Our first attempts at antigen identification utilized absorption techniques. We predicted that absorption with most antigens would cause an unavoidable absorption of protein in general; we therefore monitored the specificity of absorption with an antigen by demonstrating the following: a decrease in the IgG/albumin ratio of the specimen (indicating a preferential decline in the IgG over albumin); a loss in pre-absorption IgG bands by IEF; a recovery of the IgG bands from the absorbed pellet by elution.

We found it useful to verify the effectiveness of the technique by first studying SSPE CSF IgG. A decline in the IgG/albumin ratio was observed following absorption of SSPE CSF IgG with a pelletted measles virus preparation, but not with control Vero cells, or control brain homogenates of MBP; when SSPE CSF was tested against MS brain homogenates, the IgG/albumin ratio also decreased, perhaps because a generally large overall amount of protein was absorbed (possibly due to electrostatic interactions). Following absorption with a sufficiently large amount of virus the IgG/albumin ratio fell from .84 to .39, oligoclonal IgG bands were specifically lost from the IEF pattern and bands representing 10% of the original IgG were recovered by elution (Figure 5).

MS CSF specimens were then separately absorbed with a battery of putative antigens including: saline and acid washed control brain, saline and acid washed MS brain, MBP, crude sheep myelin, sheep purified oligodendrocytes, Vero cells, rat C6 glioma cells, human medulloblastoma cells, human red blood cells, human mononuclear cells, and torpedo electric organ. The data was difficult to interpret because several antigens, especially the brain, myelin and electric organ caused a decrease in the Ig/G albumin ratio. The changed ratio probably reflected the generally large loss of protein from these absorptions and/or binding to Fc receptors. Analysis of post-absorption IEF gels did not demonstrate any preferential loss of oligoclonal bands.

Because of difficulty in interpreting the MS absorption experiments, we decided to change our approach. It was also apparent that the presence of a large amount of nonsense antibody would make absorption with a relevant antigen difficult to detect. If one assumes that the IgG bound to MS brain (and released by acid elution) is specifically bound, then one should be able to "stick" the eluted IgG back to the brain it was eluted from. In collaboration with Dr. Joel Oger, we monitored the disappearance of IgG from (neutralized) acid eluates of the brain when recombined with the eluted brain. In order to maximize the effectiveness of absorption and to avoid antibody excess, we used iodinated IgG. As a first check on our system, we detected substantial binding of iodinated SSPE brain eluted IgG to measles virus and not Vero cells. Unfortunately, we were unable to detect binding of the SSPE IgG back to the brain from which it was eluted. The reasons for this failure may be related to denaturation of the brain measles virus antigen during elution or to insufficient antigen in the brain preparation. The lack of success with this approach in SSPE deterred us from further absorption experiments using MS brain IgG.

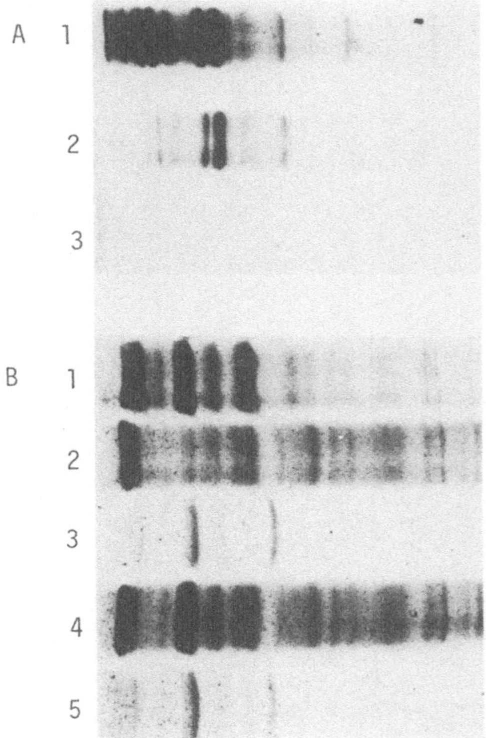

Fig. 5. Absorption of SSPE CSF oligoclonal IgG bands by measles virus. IEF pattern (with Coomassie Blue staining) of: (A) SSPE #1: CSF before absorption with measles virus (1), CSF after absorption with measles virus (2), the acid eluate of the measles pellet (3); (B) SSPE #2: CSF before absorption with measles virus (1), CSF (700 μl) after two absorptions with measles virus (2), CSF (1400 μl) after two absorptions with measles virus (4), and the acid eluates of the measles pellets in each case (3 and 5, respectively).

Immunoblot Studies

Latov et al. (1981) effectively used immunoblotting as a means of detecting the antigenic target for what must now be considered a new autoimmune, human demyelinating disease, benign monoclonal gammopathy with neuropathy (BMGN). In the immunoblotting procedure, proteins containing putative antigenic targets are subject to sodium dodecyl sulfate-polyacrylamide gel electrophoresis, transferred (blotted) to nitrocellulose paper by electroelution, overlaid with the IgG in question and then identified with an immunochemical stain. Latov et al. (1981) demonstrated that the IgM of BMGN patients immunoblots myelin associated glycoprotein (MAG).

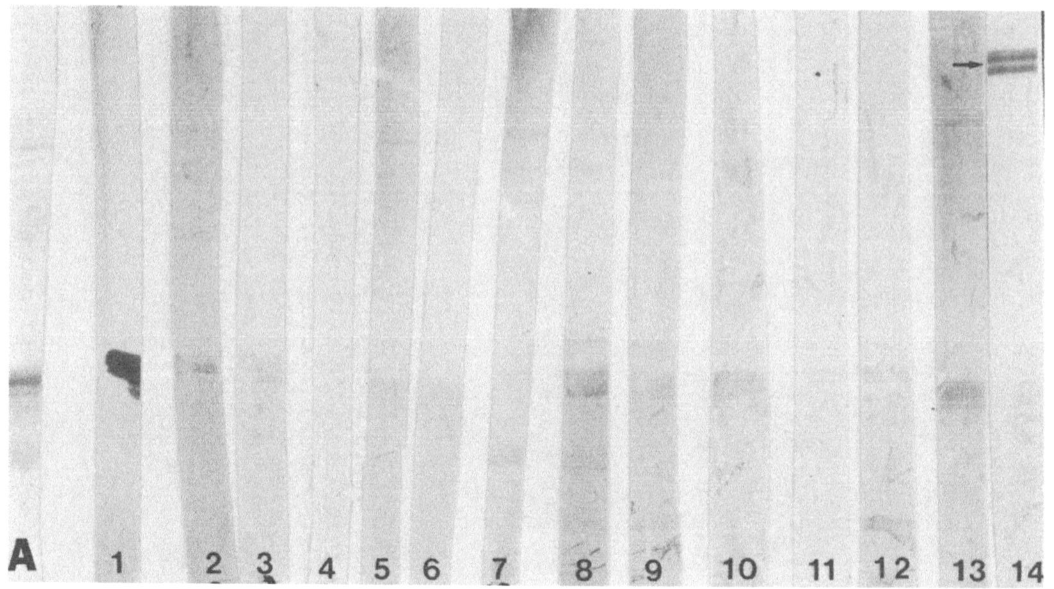

Fig. 6. Western blot in which purified myelin from MS case 1 (MS 1) was electrophoresed on a 12.5% SDS-polyacrylamide gel and then electroeluted on to nitorcellulose paper. After "quenching" of remaining binding sites with albumin, the nitrocellulose was overlaid with sera or CSF (as listed below), washed, and then overlaid with peroxidase conjugated anti-human IgM, washed and then developed with 3, 3'-diaminobenzidine tetrahydrochloride and hydrogen peroxide. Lane A is an amido black stain of the blotted proteins; the other lanes were overlaid with the following specimens: (1) normal (nl) 1 serum, (2) MS 2 serum, (3) MS 3 CSF, (5) nl 2 serum, (6) MS 1 brain eluate, (7) MS 1 serum, (8) nl 3 serum, (9) MS 5 serum, (10) MS 5 CSF, (11) MS 6 serum, (12) MS 6 CSF, (13) MS 7 CSF, (14) BMGN serum.

We have been employing this technique in order to delineate the antigenic target of MS immunoglobulin. We have used optic nerve, spinal cord, control and MS white matter, control and MS purified myelin and purified sheep oligodendrocytes (from Dr. Sara Szuchet) as antigens. We have tested these tissues with various combinations of sera and CSF from patients with optic neuritis, transverse myelitis, post-infectious encephalomyelitis and MS; IgG and IgM antibodies were screened for. Frequently, especially in overloaded gels, there was mild staining of occasional bands when either normal or abnormal sera or CSF was applied. An example of definite positive staining can be seen in Figure 5 which demonstrates MAG protein staining in a case of BMGN (Stefansson et al., 1983), and an absence of MAG protein antibody in MS CSF, sera and brain eluates. Occassional specimens had a band of immunoreactivity against CNS material, as displayed by an optic neuritis CSF in Figure 6, but generally there was little evidence of consistent reactivity. Negative immunoblot studies using MS CNS material screened against MS CSF and brain eluates have been reported by others (Newcombe et al., 1982).

Why has immunoblotting been unsuccessful? If all the MS immunoglobulin is nonsense antibody all attempts to elucidate a pathogenetic antigen will obviously fail. Our working hypothesis at present has been, however, that some portion of the MS immunoglobulin is relevant antibody. This relevant antibody may nevertheless be very scarce compared to a large amount of nonsense antibody. It may also be that MS serum and CSF contain far more nonsense antibody that brain eluates - certainly the banding patterns of MS serum, CSF and brain eluates are remarkably different (Mattson et al., 1982a; Figure 3). Perhaps eluates from MS post-mortem brain have changed over time and no longer contain antibodies that were once relevant to the disease pathogenesis. Another issue relates to the antigen we are screening for. The antigen we electrophorese may not be the correct one, or it may not be concentrated enough or it might not solubilize well or it may be denatured by SDS with a consequent loss of antigeneity. It is disquieting to realize that had we not known that acetylcholine receptor was the antigenic target of myasthenia gravis serum, we might have had a long, fruitless search screening human muscle and even endplates given the relative insolubility of the receptor and its low abundance.

CONCLUSIONS

Our studies have emphasized the differences in the MS and SSPE humoral immune response. It is clear that the MS IgG response is not as specifically directed as that seen in SSPE. The elusive antigen of the MS "orphan" antibody must await new technologies and approaches that will allow us to discriminate a reaction of a relevant specific antibody against a background of non-specific B cell responses.

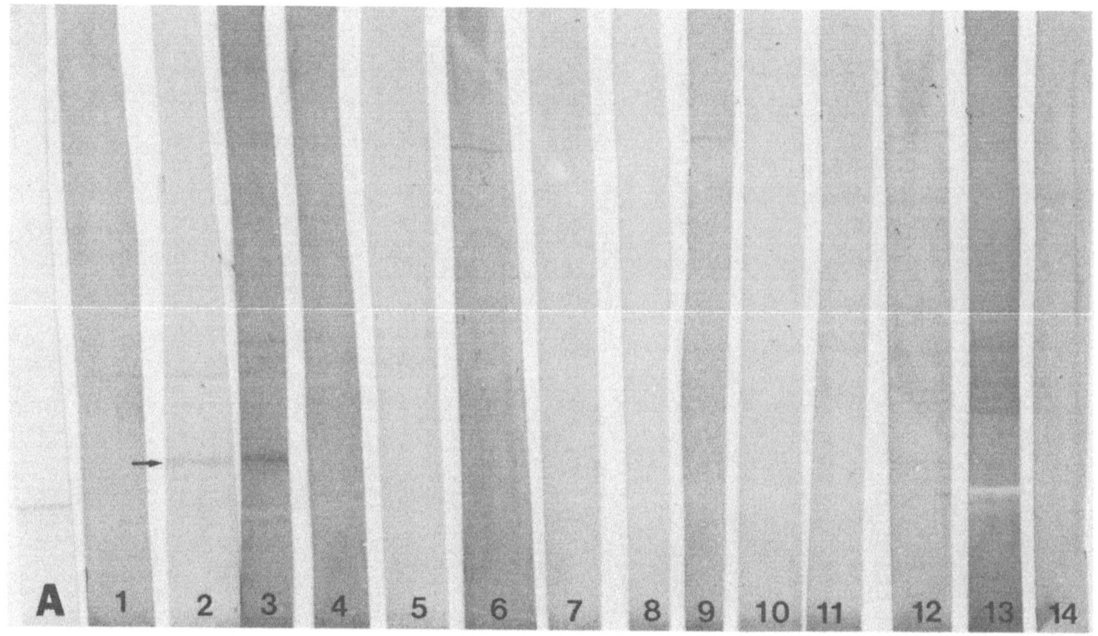

Fig. 7. Western blot in which solubilized optic nerve proteins were electrophoresed on an SDS-polyacrylamide gradient (7-25%) gel and then processed similarly to Figure 6 except peroxidase conjugated anti-human IgG rather than IgM was used. Lane A is an amido black stain of the blotted proteins. The other lanes were overlaid with the following specimens: (1) nl 1 serum, (2) optic neuritis (ON) 1 CSF, (3) ON 1 serum, (4) nl 2serum, (5) control CSF, (6) nl 3 serum, (7) ON 3 CSF, (8) ON 4 CSF, (9) nl 3 serum, (10) ON 5 CSF, (11) ON 5 serum, (12) nl 5 serum, (13) MS CSF (with optic atrophy), (14) nl 6 serum. The arrow points to a low molecular weight band stained in lane 2 and less so in lane 3. Other bands are vaguely seen and appear in some normals as well as abnormal specimens.

Acknowledgements

This work was supported by grants from the National Multiple Sclerosis Society (MS:RG 1512A-1) and the Kroc Foundation and the Medical Scientists Training Program (PHS 2-T32 GM07281 to D.H.M). The advice of Dr. Barry G. Arnason is gratefully acknowledged. We would also like to thank Zayda Stewart for excellent secretarial assistance.

REFERENCES

Barundun, S., Skragill, R., and Morell, A., (1976), Imbalance of the κ/λ of human immunoglobulins, in: "F. H. Bach and R. A. Good eds., Clinical Immunobiology, Academic Press, New York, N.Y.,

Ebers, G. C., Zabriski, J. B., and Kunkel, H. G., (1979), Oligoclonal immunoglobulins in subacute sclerosing panencephalitis and multiple sclerosis: A study of idiotypic determinants, Clin. Exp.Immunol., 35:67-75.

Gerhard, W., Taylor, A., Wroblewska, Z., et al. (1981), Analysis of a predominant immunoglobulin population in the crebrospinal fluid of a multiple sclerosis patient by means of an anti-idiotypic hybridoma antibody, Proc.Natl.Acad.Sci., USA, 78:3225-3229.

Glynn, P., Weedon, D., Edwards, J., Suckling, A. J., and Cuzner, M. L., 1982, Humoral immunity in chronic relapsing experimental autoimmune encephalomyelitis: the major oligoclonal IgG bands are antibodies to mycobacteria, J.Neurol.Sci., 57:369-384.

Karcher, D., Lassmann, H., Lowenthal, A., Kitz, K., and Wisniewski, H. M., 1982, Antibodies-restricted heterogeneity in serum and cerebrospinal fluid of chronic relapsing experimental allergic encephalomyelitis, J.Neuroimmunol., 2:93-106.

Laterre, E. C., Les proteines du liquide cephalorachidien a l'etat normal et pathologique, Brussels, Arscia, Paris, Maloine.

Latov, N., Braun, P. E., Gross, R. B., Sherman, W. H., Penn, A. S., and Chess, L., 1981, Plasma cell dyscrasia and peripheral neuropathy: identification of the myelin antigens that react with human paraproteins, P.N.A.S. 78:7139-7142.

Laurenzi, M. A., Maria, M., Kam-Hansen, S., and Link, H., Oligoclonal IgG and free light chains in multiple sclerosis demonstrated by thin-layer polyacrylamide gel isoelectric focusing and immunofixation, Ann.Neurol., 8:241-247.

Link, H., and Laurenzi, M. A., 1979, Immunoglobulin class and light chain type of oligoclonal bands in CSF in multiple sclerosis determined by agarose gel electrophoresis and immunofixation., Ann.Neurol., 6:107-110.

Lowenthal, A., van Sande, M., and Karcher, D., 1960, The differential diagnosis of neurologic diseases by fractionating electrophoretically the CSF gammaglobulins., J.Neurochem., 6:51-56.

Mattson, D. H., Roos, R. P., Hopper, J. E., and Arnason, B. G. W., 1982, Light chain composition of CSF oligoclonal IgG bands in multiple sclerosis and subacute sclerosing panencephalitis, J.Neuroimmunol., 3:63-76.

Mattson, D. H., Roos, R. P., and Arnason, B. G. W., 1980, Isoelectric focusing of IgG eluted from multiple sclerosis and subacute sclerosing panencephalitis brains, Nature, 287:335-337.

Mattson, D. H., Roos, R. P., and Arnason, B. G. W., 1982, Oligoclonal IgG in multiple sclerosis and subacute sclerosing panencephalitis brains, J.Neuroimmunol., 2:261-276.

Mattson, D. H., Roos, R. P., and Arnason, B. G. W., 1980, Immunoperoxidase staining of cerebrospinal fluid IgG in isoelectric focusing gels: a sensitive new technique, J.Neurosci. Meth., 3:67-75.

Mattson, D. H., Roos, R. P., and Arnason, B. G. W., 1981, Comparison of agar gel electrophoresis and isoelectric focusing in multiple sclerosis and subacute sclerosing panencephalitis, Ann. Neurol., 9:34-41.

Maxey, C. R., and Palmer, M. R., 1975, The behavior of gelatin preparations in the acidic region of the pH gradient, in: "Isoelectric Focusing", J. P. Arbuthnott and J. A. Beeley, eds, Butterworths, Woburn, Ma; pp.261-269.

Mehta, P. D., Miller, J. A., and Tourtellotte, W. W., 1982, Oligoclonal IgG bands in plaques from multiple sclerosis brains, Neurology 32:372-376.

Nagelkerken, L. M., Aalberse, R. C., VanWalbeek, H. K. V., et al., 1980, Preparation of antisera directed against the idiotype(s) of immunoglobulin G from the cerebrospinal fluid of patients with multiple sclerosis, J.Immunol., 125:384-389.

Newcombe, J., Glynn, P., and Cuzner, M. L., 1982, The immunological identification of brain on cellulose nitrate in human demyelinating disease, J.Neurochem., 38:267-274.

Newcombe, J., Glynn, P., and Cuzner, M. L., 1983, Analysis by transfer electrophoresis of reactivity of IgG with brain proteins in multiple sclerosis, J.Neurochem., 39:1192-1194.

Norrby, E., Link, H., Olsson, J.-E., Panelius, M., Salmi, A., and Vandvik, B, 1974, Comparison of antibodies against different viruses in cerebrospinal fluid and serum samples from patients with multiple sclerosis, Infect.Immun., 10:688-694.

Oger, J. J.-F., Mattson, D., Roos, R., Antel, J., and Arnason, B. G. W., 1981, Isoelectric focusing of IgG secreted by human peripheral lymphocytes in multiple sclerosis and controls, Neurology., 21:144.

Roos, R., and Lichter, M., Silver staining of cerebrospinal fluid IgG in isoelectric focusing gels, J.Neuroscience Methods, In press.

Rostrom, B., Link, H., Laurenzi, M. A., Kam-Hansen, S., Norrby, E., and Wahren, B., 1981, Viral antibody activity of oligoclonal and polyclonal immunoglobulins synthesized within the central nervous system in multiple sclerosis, Ann.Neurol., 9:569-574.

Siemes, H., Siegart, M., Hanefeld, F., Kolmel, H. W., and Paul, F., 1977, Oligoclonal gamma-globulin banding of cerebrospinal fluid in patients with subacute sclerosing panencephalitis, J.Neurol. Sci., 32:395-409

Stefansson, K., Marton, L., Antel, J. P., Wollmann, R. L., Roos, R. P., Chejfec, C., and Arnason, B. G. W., 1983, Neuropathy accompanying IgMλ monoclonal gammopathy, Acta.Neuropathologica, 59:255-000.

Tachovsky, T. G., Sandberg-Wolheim, M., and Baird, L. G., 1982, CSF igG characterization of antidiotype antibodies produced against MS CSF and detection of cross reactive idiotypes in several MS CSF, J.Immunol., 129:764-770.

Vandvik, B., Nilsen, R. E., Vartdal, F., and Norrby, E., Mumps meningitis, specific and non-specific antibody responses in the central nervous system, Acta Neurol. Scand., 65:468-487.

Vartdal, F., and Vandvik, B., 1982, Multiple sclerosis: electrofocused "bands" of oligoclonal CSF IgG do not carry antibody activity agains measles, varicella-zoster or rotaviruses, J.Neurol.Sci., 54:99-107.

Whitacre, C. C., Mattson, D. H., Day, E. D., Peterson, D. J., Paterson, P. Y., Roos, R. P., and Arnason, B. G. W., 1982, Oligoclonal IgG in rabbits with experimental allergic encephalomyelitis: non-reactivity of the bands with sensitizing neural antigens, Neurochem.Res., 7:1209-1221.

Whitacre, C. C., Mattson, D. H., Paterson, P. Y., Roos, R. P., Peterson, D. J., and Arnason, B. G. W., 1981, Cerebrospinal fluid and serum oligoclonal IgG bands in rabbits with experimental allergic encephalomyelitis, Neurochem.Res., 6:87-96.

STUDIES ON B-LYMPHOCYTE FUNCTION IN MULTIPLE SCLEROSIS

Hans Link[1], Slavenka Kam-Hansen[1] and Annemarie Henriksson[2]

[1]Department of Neurology, Karolinska Institutet, Huddinge University Hospital, S-141 86 Huddinge, Stockholm and
[2]Department of Neurology, University Hospital, S-581 85 Linköping, Sweden

INTRODUCTION

Among abnormalities reported in multiple sclerosis (MS) and considered to reflect derangements of the immune system, those of B-lymphocyte function within the CNS-CSF compartment are generally accepted as the most consistent and clear-cut. These abnormalities have recently been extensively reviewed (Trotter and Brooks, 1980; Brooks et al., 1983; Walsh and Tourtellotte, 1983; Walsh et al., 1983). They include elevation of the CSF-IgG/total protein ratio (Kabat et al., 1948) and the CSF-IgG index, equal to (CSF-IgG/serum-IgG): (CSF-albumin/serum-albumin) (Tibbling et al., 1977), accentuated predominance in CSF of IgG1 (Palmer and Minard, 1976) and of IgG kappa (Link and Zettervall, 1970), and oligoclonal distribution of IgG on separation of CSF by electrophoresis or isoelectric focusing on suitable media such as agar or agarose, in about 90%. MS patients heterozygous to G1m (1) and G1m (3) allotypes of IgG1 showed greatly increased concentrations in the CSF of G1m[1] in comparison with patients with other neurological disorders indicating that, in the heterozygous MS patient, most plasma cells in the CNS-CSF compartment preferentially synthesize G1m (1) IgG1 molecules (Salier et al., 1981). Furthermore, unexpected Gm allotypes may be found in CSF but not serum from an occasional MS patient (Salier et al., 1983). A majority of patients with MS display evidence for intrathecal synthesis of IgG class antibodies against various viruses, especially measles (for review see Norrby, 1978). Simultaneous intrathecal production of antibodies against 11 different viral antigens in individual MS patients has been reported (Salmi et al., 1983). Furthermore, synthesis may occur within the CNS-CSF compartment of antibodies against bacteria (Vartdal et al., 1980) and various CNS epitopes (for review see Roström, 1981; Walsh et al., 1983).

Recently, elevated CSF-IgM index reflecting intrathecal IgM production has been reported by others (Sindic et al., 1982) and by us (Forsberg et al., 1983) in about 60% of patients with MS (Figure 1). A parallel may be drawn with the situation in neurosyphilis where a prolonged intrathecal IgM-response has been described (Strandberg-Pedersen et al., 1982), which may normalize several months after successful treatment of the disease. The antibody specificity(ies) of locally produced IgM in MS remain to be elucidated.

The above abnormalities are all consistent with B-lymphocyte hyperfunction and the consequence of increased concentrations of free immunoglobulins (Igs) within the CNS-CSF compartment. Whether aberrations of free Ig measured in the form of these variables are really representive for existing derangements of B-lymphocyte functions is, however, far from clear. This might be elucidated by studying Igs in supernatants of cultivated CSF mononuclear cells. In such a way, " a complete repertoir" or Ig-producing cells from CSF compartment could be obtained.

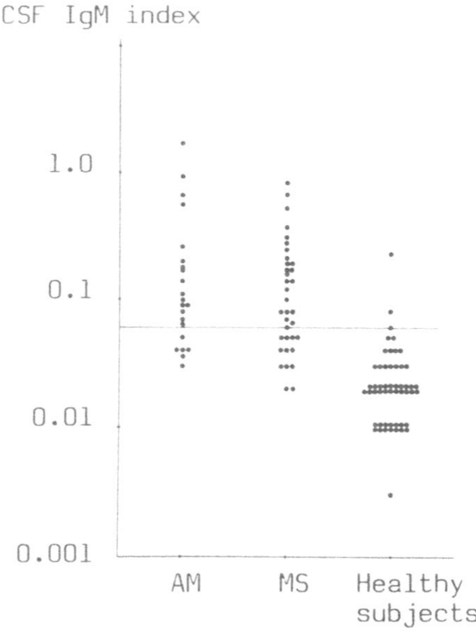

Fig. 1. CSF-IgM index (logarithmic scale) for reference group of 51 "healthy subjects", i.e. individuals mostly with tension headache and with normal findings at neurological and routine CSF examinations, and for 35 patients with multiple sclerosis (MS) and 22 with aseptic meningoencephalitis (AM). Our upper reference limit (mean + 2SD) for CSF-IgM index of 0.06 is drawn.

The concentrations in CSF of individual Igs as well as antibodies can be considered to be determined by an afferent and an efferent axis.

Afferent axis	Efferent axis
Transudation via blood-brain barrier	Metabolism
Local production within CNS and meninges	Differences in half-life (IgG 25 days; IgA 6 days; IgM 5 days) (Hood et al., 1978)
Local synthesis by CSF lymphocytes	Proteolysis
	Consumption at target

Local synthesis by lymphocytes present in the CSF is probably quantitatively of low-grade importance in comparison with the two other mechanisms constituting the afferent axis. The mechanisms of the efferent side have hitherto mostly been neglected in discussions of Ig concentrations in CSF. Greatly elevated activities of the peptidase leucyl-B-naphthylamide (Rinne and Riekkinen, 1968) and of acid and neutral proteinases (Rinne and Riekkinen, 1968; Cuzner et al., 1979) have been reported in CSF from patients with MS. To what extent these abnormalities influence Ig levels in CSF is not known. The hitherto unknown relation between pentamers and monomers of IgM in CSF should have influence on IgM concentrations in CSF. Little is known about consumption of antibodies at targets but it may be anticipated that his mechanism has substantial influence on concentrations in CSF of Igs of various classes and subclasses, and of antibody specificities.

Concentrations of free Igs in water-soluble CNS extracts are probably influenced by similar mechanisms. Consumption at target can be exemplified by our observation that water-soluble CNS extracts obtained from guinea pigs with chronic relapsing experimental allergic encephalomyelitis (r-EAE) did not contain measurable amounts of IgG antibodies against myelin when examined by a sensitive enzyme-linked immunosorbent assay (ELISA) while the same assay revealed high concentrations of such antibodies in supernatants from short-term (18 h) cultures of CNS and meningeal lymphocytes obtained from the same animals (Olsson et al., manuscript in preparation).

It can be anticipated that studies performed directly on CSF-lymphocytes (CSF-L) in comparison with peripheral blood lymphocytes (PBL) from the same individual will give information in addition to that available from determination of concentrations of free Igs or antibodies in CSF. In this paper, we present results from our laboratory regarding enumeration of B-lymphocytes in CSF and the peripheral blood; enumeration of Ig-secreting cells in CSF and peripheral blood by applying an indirect haemolytic plaque-forming cell (PFC) assay; and analysis of concentrations of Igs secreted into supernatants obtained from long-term (7 days) cultures of CSF-L and PBL obtained from patients with MS and from controls. Our results indicate that enumeration of PFC and determination of concentrations of Igs - and probably of antibodies - in lymphocyte culture supernatants reflect B-lymphocyte hyperfunction more appropriately than concentrations of free Igs in body fluids.

B-LYMPHOCYTES IN CSF AND PERIPHERAL BLOOD IN MULTIPLE SCLEROSIS

On stimulation, B-lymphocytes are transformed from the resting state to plasma cells which secrete Ig belonging to different classes and subclasses, and carrying various antibody specificities. B-lymphocytes have markers consisting of Ig molecules on the surface membrane, while plasma cells lack such markers.

The topic on B-lymphocyte proportions in CSF and peripheral blood in MS and neurological disorders has been carefully reviewed (Kam-Hansen, 1980; Brooks et al., 1983; Walsh et al., 1983). Great controversy exists on the presence of consistent differences in lymphocyte populations, including B-cells, in patients with MS. We found significantly ($p < 0.001$; Student's t-test) lower mean B-lymphocyte value in peripheral blood from a group of 18 MS patients compared to patients with mumps meningitis as well as healthy controls (Figure 2). All patients with MS had a pleocytosis and oligoclonal IgG in the CSF, and 14 of the patients had elevated CSF- IgG index. The mean B-lymphocyte value was slightly lower in CSF compared to peripheral blood in the MS patient group (Kam-Hansen et al., 1978). These data are consistent with our subsequent finding in MS patients (vide infra) of increased numbers, especially in CSF but also in peripheral blood, of Ig-secreting cells lacking surface Ig and thus no longer identifiable as B-lymphocytes. It must, on the other hand, be emphasized that quantitation of B- and of total T-lymphocytes comprises a very gross assessment of lymphocyte populations, and more sophisticated tests are warranted to detect effects of possible derangements in immune regulation in MS.

IMMUNOGLOBULIN-SECRETING CELLS IN CSF AND PERIPHERAL BLOOD IN MULTIPLE SCLEROSIS AND CONTROLS

Enumeration of mature B-lymphocytes, i.e. plasma cells secreting Ig, in CSF and peripheral blood from patients with MS has given us a completely different picture than mere quantitation of B-lymphocytes. Enumeration of Ig-secreting cells may be carried out by an indirect haemolytic PFC assay which is based on the capacity of one single

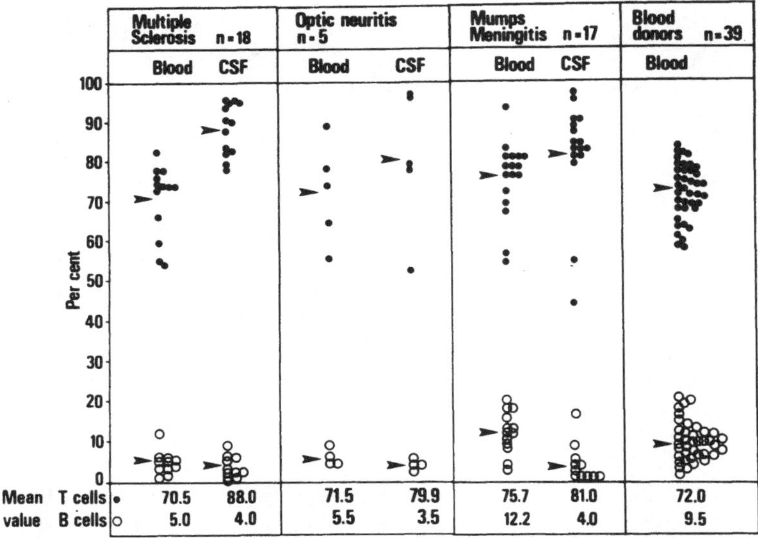

Fig. 2. Percentages of B cells (o) and T cells (•) in blood and cerebrospinal fluid (CSF) from patients with multiple sclerosis, optic neuritis and mumps meningitis, and in blood only from blood donors. Arrows depict mean values.

plasma cell to produce and secrete Ig molecules of a certain class and antibody specificity. In this assay (Figure 3), secretion is detected in a system comprising sheep red blood cells (SRBC) coated with protein-A which binds Fc part of antibody molecule directed against heavy chains of the human Ig class to be detected. The antibody molecules bind Ig secreted by plasma cells. In presence of complement, lysis of SRBC is induced and this haemolysis is then used as indicator of secretion. Lymphocytes are prepared from peripheral blood by Ficoll-Isopaque centrifugation, and from CSF by centrifugation at 200 x g for 10 min at 4°C. The indirect haemolytic PFC assay has been adopted in our laboratory for the study of 20 x 10^3 lymphocytes, in exceptional cases even 10 x 10^3 lymphocytes (Henriksson et al., 1981). This makes it possible to examine CSF specimens with slight pleocytosis or even normal mononuclear cell counts. Lymphocytes, protein-A-coated SRBC and monospecific antiserum are mixed together with complement and agar, and poured on a Petri dish. After incubation at 37°C for 6 h, haemolytic plaques are counted visually. Each plaque corresponds to one Ig-secreting cell. In order to establish a safe marginal, the occurrence of one or two plaques was judged as negative.

In peripheral blood from healthy subjects, we found a predominance of IgA-producing cells, followed by cells producing IgG and IgM (Figure 4). In normal CSF, it is difficult to enumerate Ig-producing

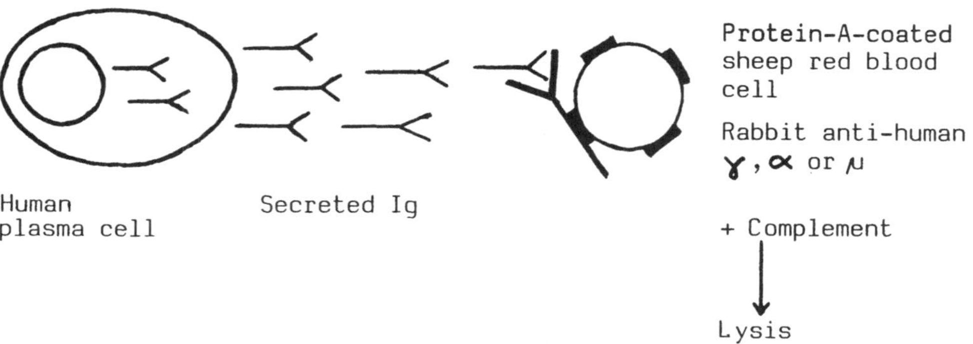

Fig. 3. Principles of indirect haemolytic plaque-forming cell (PFC) assay.

cells because of the cell paucity. In a study of four individuals with tension headache, no Ig-producing cells per 10 x 10^3 lymphocytes used in the test were detectable in the CSF of these presumably healthy individuals. For these reasons, we consider the presence of more than two plaques on examination of 20 x 10^3 CSF lymphocytes as evidence of B-lymphocyte activation in the CSF compartment.

Enumeration of PFC in CSF and peripheral blood from 11 patients with MS (Figure 4) revealed that 10 of the patients had higher numbers of IgG-producing cells per 20 x 10^3 lymphocytes in CSF than in peripheral blood (Henriksson et al., 1981). These are expected results based on our previous knowledge of synthesis of free IgG within the CNS in most patients with MS (see Introduction). Interestingly, 9 of the 11 patients had IgA-producing cells in CSF and 7 had IgM-producing cells, indicating intrathecal production also of Igs belonging to these two classes in the majority of MS patients. There were, however, no significant differences in these 11 MS patients between CSF and peripheral blood for numbers of cells secreting IgA or IgM.

To allow analysis of numbers of Ig-producing cells inter alia in relation to clinical variables, the indirect haemolytic PFC assay was applied on a larger material consisting of 37 ambulatory patients with MS (Henriksson et al., manuscript in preparation). None of these patients had been treated with immunosuppressive or immunostimulatory drugs for the last year, and the majority had never received such medications. All patients had oligoclonal IgG in CSF.

Among these 37 patients with MS, 89% had cells in CSF producing IgG, 70% had cells producing IgM and 57% had cells producing IgA. No patient was found who was negative in CSF for Ig-producing cells. Interestingly, four of the patients had in their CSF IgM-and/or IgA-producing cells but no cells producing IgG.

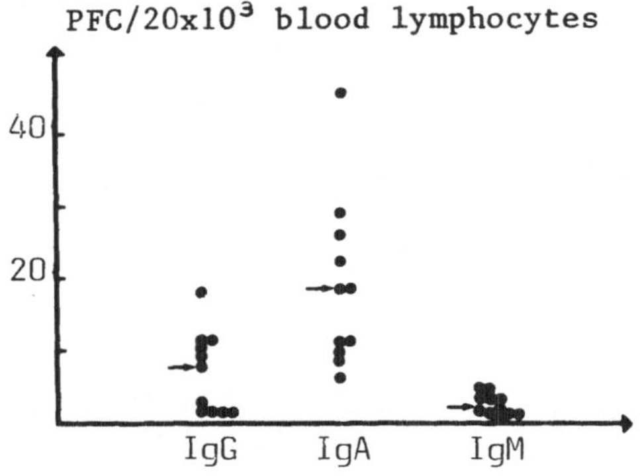

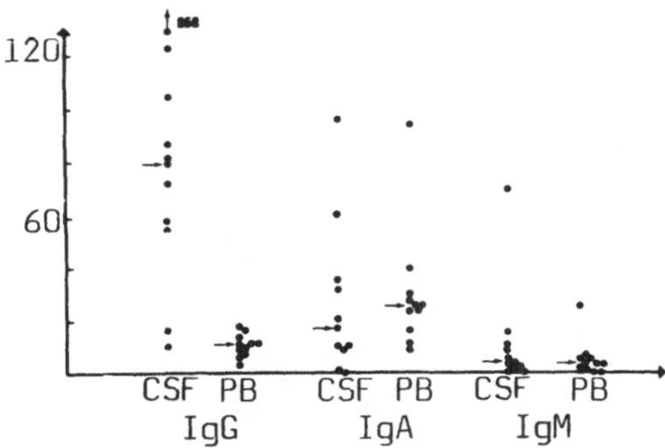

Fig. 4. Distribution of Ig-producing cells among 20 x 10^3 lymphocytes from peripheral blood from 11 healthy controls (upper part) and from CSF and peripheral blood (PB) from 11 patients with MS (lower part). Arrows denote median values.

Our previous finding of much higher numbers of IgG-producing cells in CSF than peripheral blood was confirmed on this larger material of 37 patients with MS (Table 1). In contrast, the number of IgA-producing cells was significantly lower in CSF, while the number of IgM-producing cells did not differ in CSF and peripheral blood.

Among these 37 patients with MS, IgG-producing cells predominated in CSF in 27, IgM-producing cells in 6, and IgA-producing cells in 3, while one patient had the same numbers of IgG- and IgA- producing cells in CSF (Table 2). In contrast, IgA-producing cells predominated in peripheral blood in most of the patients, as in the healthy controls. Among the 6 patients in whom IgM-producing cells predominated in CSF, two had active and four stable MS. Among those two in whom IgM-secreting cells predominated in peripheral blood, one had active and one stable MS.

Table 1. Absolute Numbers of Plaque Forming Cells per 20 x 10^3 Lymphocytes in Peripheral Blood and CSF from 37 Patients with Multiple Sclerosis (MS) and After Subgrouping in Active and Stable MS, and 20 Patients with Acute Aseptic Meningo-Encephalitis (AM) Examined within 10 Days After onset, and in Peripheral Blood only from 27 Healthy Controls.

Diagnosis and numbers of subjects		IgG		IgA		IgM		Sum of IgG + IgA + IgM	
		Blood	CSF	Blood	CSF	Blood	CSF	Blood	CSF
MS (n=37)	Range	2-26	0-858	0-13	0-94	0-27	0-307	7-169	6-989
	Median value	10	89	25	10	4	4	38	108
Active MS (n=13)	Range	2-19	0-858	8-95	0-60	0-27	0-77	12-141	15-989
	Median value	8	89	17	5	4	4	30	108
Stable MS (n=24)	Range	2-26	0-520	0-135	0-94	0-23	0-307	7-169	6-553
	Median value	10	91	25	10	4	3.5	40.5	110
AM (n=20)	Range	0-319	0-716	7-295	0-596	4-99	0-326	21-482	4-1508
	Median value	31	16	83	29	17	9	129	74
Healthy controls (n=27)	Range	2-20	n.d.	4-42	n.d.	1-7	n.d.	8-68	n.d.
	Median value	7		12		2		21	

n.d. = not done

Table 2. Distribution of Predominating Class Specificity of Ig-Producing Cells in Peripheral Blood and CSF From 37 Patients with Multiple Sclerosis (MS) and After Subgrouping in Active and Stable MS, and 20 Patients with Acute Aseptic Meningo-Encephalitis (AM) Examined Within 10 Days After Onset, and in Peripheral Blood Only From 27 Healthy Controls.

Diagnosis and numbers of Subjects	IgG		IgA		IgM		IgG = IgA		IgG = IgM	
	Blood	CSF	Blood	CSF	Blood	CSF	Blood	CSF	Blood	CSF
MS (n=37)	3	27	32	3	2	6	0	1	0	0
Active MS (n=13)	1	11	11	0	1	2	0	0	0	0
Stable MS (n=24)	2	16	21	3	1	4	0	1	0	0
AM (n=20)	3	8	15	9	2	2	0	0	0	1
Healthy controls	2	n.d.	24	n.d.	0	n.d.	1	n.d.	0	n.d.

n.d. = not done

A positive correlation was found among the 37 MS patients between the numbers of IgM-producing cells in CSF and the CSF-IgM index. No such correlations were found for IgG or IgA. For peripheral blood, positive correlations were registered between serum IgG concentrations and numbers of IgG-producing cells, and also between serum IgA concentrations and numbers of IgA-producing cells, while no correlation was found concerning IgM.

Altogether 28 of the 37 MS patients had in their CSF cells producing Ig of a certain class when the corresponding CSF-Ig index was normal (Table 3). These data confirmed that enumeration of Ig-PFC is a more sensitive variable of the intrathecal immune status than determination of free Ig concentrations, even when presented as corresponding CSF-Ig index.

In peripheral blood, our 37 patients with MS had significantly higher numbers of IgG-, IgA- and also IgM-producing cells compared with the 27 health controls. These findings indicate that MS is also characterized by a systemic B-lymphocyte hyperactivity which is not obvious when free serum Ig concentrations are determined.

It was possible to subgroup the 37 patients with MS into 13 with active disease, i.e. clinical relapse or progressive disease for more than 6 months duration, and the remaining 24 patients with stable disease defined as no symptoms occurring over an interval extending from 3 months before and 1.5 months after time of testing. No sig-

Table 3. Relation Between Numbers of Ig-Producing Plaque-Forming Cells (PFC) per 20 x 10^3 CSF Lymphocytes and Corresponding CSF-Ig Index in 37 Patients with Multiple Sclerosis

	Normal	Increased
CSF-IgG index (n=37)		
IgG PFC		
Normal ($\leqslant 0.7$)	0	5
Increased (> 0.7)	4	28
CSF-IgA index (n=31)		
IgA PFC		
Normal ($\leqslant 0.6$)	8	22
Increased (> 0.6)	0	1
CSF-IgM index (n=35)		
IgM PFC		
Normal ($\leqslant 0.06$)	10	4
Increased (> 0.06)	5	16

nificant differences for Ig-producing cells were found between the two groups (Table 1). All those abnormalities encountered for the whole material of 37 patients were also found in both groups, i.e. increased numbers of IgG-producing cells and decreased numbers of IgA-producing cells in CSF, and no significant difference for IgM-producing cells in CSF compared to peripheral blood.

Acute aseptic meningo-encephalitis (AM) is a benign self-limiting viral infection of the CNS which may be accompanied by elevated concentrations in CSF of IgG, IgA and/or IgM (Frydén et al., 1978) and, in about a third of the cases, by oligoclonal IgG bands in CSF (Laterre et al., 1970; Link and Müller, 1971). The mononuclear pleocytosis encountered in this group of patients makes them suitable as controls for studies of intrathecal immune reactions in MS. We have found that in acute AM, as a reflection of generalized immune system stimulation, there is a predominance of Ig-producing cells in peripheral blood compared with in CSF (Table 1), although the reverse was found in individual patients (Henriksson et al., 1981); Forsberg and Kam-Hansen, 1983). Positive correlations were observed in the acute stage of AM between numbers of Ig-producing cells in CSF of all three classes and the corresponding CSF-Ig index values, while no positive correlations were observed between the numbers of IgG-, IgA- or IgM-producing cells in peripheral blood and the corresponding serum Ig concentrations. Since bone marrow - but not the blood - is the major source of the Igs in serum (Hijmans and Schuit, 1972), no such correlations are to be found in healthy subjects, while in paraproteinemias positive correlations occurred. The positive correlations found in peripheral blood from our MS patients for IgG and IgA may reflect a situation comparable to a low-grade IgG and IgA paraproteinemia.

Several of the patients with AM had elevated numbers of Ig-producing cells in the CSF in the presence of normal values of the corresponding CSF - Ig index. These observations also suggest that enumeration of Ig-producing cells in CSF is a more sensitive indicator of the immune status intrathecally than measurement of the free Igs. The sensitivity of the indirect haemolytic PFC assay is also corroborated by the finding that a few patients with AM during convalescence had normal leucocyte counts per microlitre of CSF, while at the same time considerable proportions of the CSF cells were still active Ig producers.

In conclusion, our studies applying an indirect haemolytic PFC assay to enumerate Ig- (antibody-) producing cells in CSF in MS have confirmed the occurrence of intrathecal B-lymphocyte hyperactivity: Ig-producing cells were regularly demonstrable in MS-CSF, often in high numbers and mostly secreting IgG; a pronounced intrathecal IgM response was found, reflected by presence of IgM-producing cells in 70% of our 37 MS patients and, in fact, a predominance of IgM-producing cells in 6 (16%) of these patients; finally, 28 of these 37 patients had in CSF cells producing Igs of G, A or M class in presence of normal values of corresponding CSF-Ig index, confirming that enumeration of Ig-producing cells in CSF is a more sensitive indicator of the intrathecal immune status than measurements of concentrations of free Igs. Regarding peripheral blood, systemic B-lymphocyte hyperactivity has been demonstrated in MS in the form of increased numbers of Ig-producing cells in presence of normal serum concentrations of free Igs.

Future perspectives opened by application of an indirect haemolytic PFC assay in MS include definition of the humoral immune response on the IgG subclass level in more detail. There may be not only an increase of IgG1 but also derangements in proportions between IgG or various subclasses having aggressive and protective effects, respectively. Modification of the assay to allow enumeration of cells in CSF and peripheral blood producing specific antibodies has high priority. Finally, it should be of interest to examine numbers of Ig-producing cells in peripheral blood and CNS in relation to concentrations of free Igs in r-EAE. In this experimental counterpart to MS, our data point to several similarities when B-lymphocyte function is considered (Table 4).

IN VITRO SYNTHESIS OF IMMUNOGLOBULINS BY CEREBROSPINAL FLUID AND PERIPHERAL BLOOD CELLS IN MULTIPLE SCLEROSIS AND CONTROLS

CSF mononuclear cells from patients with MS have been shown to have the capacity to produce IgG and also IgA when cultured in vitro

Table 4. Similarities of B-Lymphocyte Function Between Multiple Sclerosis and Chronic Relapsing Experimental Allergic Encephalomyelitis (r-EAE).

1. Elevated concentrations of IgG and IgM in CSF and CNS
2. Elevated concentrations of serum IgG antibodies against myelin and myelin basic protein
3. Oligoclonal IgG present in serum, CSF and CNS
4. Patterns of Oligoclonal IgG differ in various regions of individual CNS
5. Oligoclonal IgG does <u>not</u> contain significant amounts of antibodies against myelin, myelin basic protein or galactocerebroside
6. The major portion of oligoclonal IgG in CSF and CNS has not been defined with regard to antibody specificity
7. No relations between Ig levels or oligoclonal IgG, and exacerbations or remissions

(Sandberg-Wollheim, 1974). The synthesized IgG had an oligoclonal distribution and showed the same electrophoretic pattern as the IgG in the CSF. Blood lymphocytes from the same patients were found to synthesize an IgG in vitro that showed a completely different electrophoretic pattern. The question about occurrence of oligoclonal IgG bands in MS patients' serum has later been elaborated with more sensitive protein separation techniques, mainly isoelectric focusing on polyacrylamide gels. Using subsequent immunofixation with monospecific antisera, Laurenzi et al., (1980) showed that no less than 74% of MS patients displayed oligoclonal IgG bands simultaneously in CSF and in serum. Even though the bands had identical migration properties in CSF as in serum, those in serum were less in number and also weaker. These observations were interpreted as a result of intrathecal IgG production and subsequent transudation to serum, or occurrence of the same lymphocyte clones producing identical Ig inside and outside the blood-brain barrier. Contradictory to these findings, Mehta et al., (1982) reported oligoclonal bands in about 50% of MS sera with patterns either only partially similar or completely different from those seen in matching CSF. Whether this is a consequence of isolation of serum-IgG carried out by these authors prior to isoelectrofocusing or really reflects extrathecal B-cell hyperfunction remains to be settled. The question might be resolved by repeating the experiments of Sandberg-Wollheim (1974), but with the use of more sensitive IgG separation and detection techniques such as isoelectric focusing followed by double antiserum-biotin-avidin-peroxidase staining (Olsson et al., 1983) of separated oligoclonal IgG.

At the time when Sandberg-Wollheim (1974) performed her studies, less sensitive methods were available for e.g. quantification of Igs and this may be a reason why IgM was not demonstrable in the culture supernatants. Furthermore, a minimum of 0.25×10^6 CSF-L and 10^7 PBL were needed for each in vitro culture.

On the basis of our findings of elevated CSF - IgM index and of IgM- and IgA-producing cells in CSF in a majority of patients with MS, reflecting intrathecal production of these Igs, we analyzed culture supernatants obtained from CSF -L and PBL cultivated for 7 days without and in presence of pokeweed mitogen (PWM), for concentrations of Igs belonging to the three main classes (Henriksson et al., manuscript in preparation). PWM is a mitogen which stimulates B-cells, but also T-cells. The procedure of lymphocyte culturing and determination of Ig concentrations in culture supernatants is summarized in Figure 5.

The supernatants from unstimulated cultures of PBL from 11 healthy controls revealed the highest values for IgM concentrations, followed by IgA and IgG (Table 5). Cultures in presence of PWM resulted in increases of concentrations of Ig of all three classes in the supernatants, being about 4-fold for IgM, 13-fold for IgA and

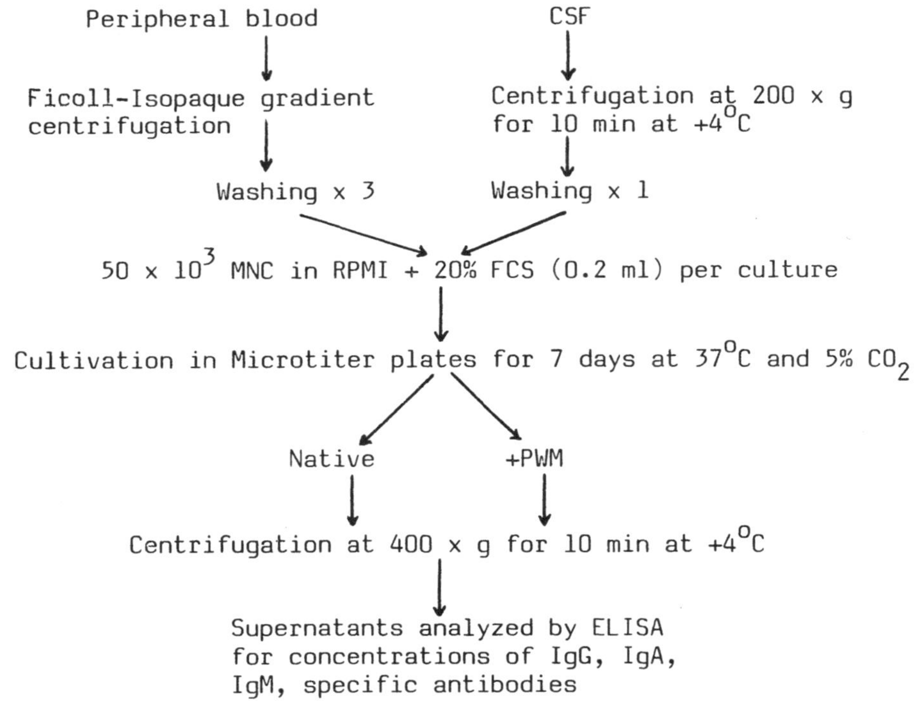

Fig. 5. Scheme of procedure used for mononuclear cell (MNC) separation and cultivation, and determination of Ig concentrations in culture supernatants.

33-fold for IgG. The differences between values of PWM-stimulated and unstimulated cultures can be considered to represent the net potential of stimulated lymphocytes to produce Ig. This difference was highest for IgM, followed by IgA and IgG among the healthy controls.

In 11 MS patients higher IgG concentrations were found in supernatants from unstimulated PBL cultures when compared with the corresponding values in healthy controls (Table 5). For IgM and IgA, no such differences were observed. PWM induced significant increase of concentrations of Ig belonging to all three classes, being about 15-fold for IgM, 13-fold for IgA and 9-fold for IgG. The difference between values of PWM-stimulated and unstimulated PBL cultures - reflecting the net potential of stimulated lymphocytes to produce Ig - was highest for IgM, followed by IgG and IgA (Table 5).

Goust et al., (1982), studying PBL only, have shown that PBL from MS patients with active disease in 7 days cultures produced significantly more IgG after stimulation with PWM in vitro than PBL

Table 5. Median Values of Ig Concentrations in Supernatants from Unstimulated and PWM-Stimulated Cultures of Lymphocytes From Peripheral Blood and CSF from Patients with Multiple Sclerosis (MS) and Acute Aseptic Meningo-Encephalitis (AM), and from Peripheral Blood in Healthy Controls.

Diagnosis and number of subjects	PWM	IgG		IgA		IgM	
		Blood	CSF	Blood	CSF	Blood	CSF
MS	-	292	904	120	25	343	248
(n=11)	+	2 672	743	1 567	31	5 283	48
	Difference[1]	2 004	9	1 307	5	4 316	4
AM	-	163	85	84	20	517	51
(n=11)	+	2 800	640	943	83	5 040	248
	Difference[1]	1 578	610	770	62	4 918	87
Healthy subjects	-	40	n.d.	108	n.d.	333	n.d.
(n=11)	+	1 326	n.d.	1 370	n.d.	1 454	n.d.
	Difference[1]	1 282		1 343		2 693	

[1]Difference means median value of (Ig concentration in supernatants from PWM-stimulated cultures - Ig concentration in corresponding supernatants from unstimulated cultures), equal to net potential of stimulated lymphocytes to produce Ig.
n.g. = nod done

from normal age-matched controls. Our data are thus in agreement with those of Goust et al., and represent an extension with regard to IgA and IgM.

Supernatants of unstimulated CSF-L cultures from the MS patients contained higher IgG concentrations in comparison with supernatants of corresponding unstimulated PBL cultures. Although much higher concentrations of both IgA and IgM were found in supernatants from unstimulated PBL than CSF-L suprnatants, illustrating the intensity of intrathecal IgG response in MS. As for PBL, IgM concentrations in unstimulated CSF supernatants were higher than those of IgA.

In strong contrast to the situation for PBL, CSF-L from the MS patients did not respond to PWM. Thus, the concentrations in the CSF-L supernatants of Ig belonging to all three classes remained unaffected by PWM, which means that the strong IgG predominance from unstimulated culture supernatants persisted.

Previously, low or absent proliferative response of MS-CSF-L to PWM has been reported (Kam-Hansen et al., 1979). Low or absent responses of MS-CSF-L have also been obtained with two other mitogens, namely phytohaemagglutinin and concanavalin A. One explanation for this poor response to mitogens might be that the MS-CSF-L are already maximally activated, according to one hitherto unproven proposal because of viral infection of the lymphoid cells (Kam-Hansen, 1980).

In 11 AM patients examined as controls, higher concentrations of lgM and IgG were found in supernatants of unstimulated PBL when compared with healthy subjects, and of IgM only when compared with MS patients, while no such differences were registered for IgA. PWM induced increased concentrations of the Ig belonging to all three classes in the PBL supernatants, being about 10-fold for IgM, 11-fold for IgG and 17-fold for IgG. THe difference between Ig concentrations in PWM-stimulated and unstimulated PBL cultures, reflecting the net potential of stimulated lymphocytes to synthesize Ig, was highest for IgM, followed by IgG and IgA, as in MS. The values were somewhat higher in MS for IgA and IgG.

Supernatants of unstimulated CSF lymphocytes displayed much lower concentrations of the three Ig classes when compared with the corresponding PBL. This is in contrast to the situation in MS for IgG which predominated markedly in CSF lymphocyte supernatants. Even the sum of concentrations of all three Igs in the CSF lymphocyte supernatants of the AM patients was only about 20% of that in corresponding PBL supernatants. While the concentrations of IgM and IgA were similar in unstimulated CSF lymphocyte supernatants of patients with MS and AM, the IgG concentration was much lower in AM.

Significant increases of concentrations of the three Igs were induced by PWM-stimulation of CSF lymphocytes in AM, being about 5-fold for IgM, 4-fold for IgA and 8-fold for IgG. This is in sharp contrast to MS CSF-L where no significant increase in concentration could be demonstrated for any of the three Igs. The difference between Ig concentrations in PWm-stimulated and unstimulated CSF lymphocyte cultures, i.e. the net potential of stimulated CSF lymphocytes to produce Ig, was greatest for IgG, followed by IgM and IgA.

Summarizing these data from the studies on concentrations of IgG, IgA and IgM in lymphocyte culture supernatants, it has become clear that the IgG concentration in unstimulated MS CSF-L culture supernatants was about three times higher than in the corresponding PBL culture supernatants. PWM induced no increase of IgG, nor of IgA or IgM in MS CSF-L supernatants, in contrast to the situation in AM. PBL from the MS patients responded to PWM as did PBL from healthy subjects and from patients with AM. One explanation for the unresponsiveness of MS CSF-L might be that they are already maximally activated. The results from the supernatant studies also confirmed that CSF mononuclear cells from most MS patients have the capacity to produce IgG as well as IgA and IgM, and that PBL are activated to a higher degree in MS than in healthy subjects regarding IgG production.

CONCLUDING REMARKS

MS is characterized by intrathecal activation of B-lymphocytes, the cause of which is still unsettled. It might be the result of specific stimulation with a defined antigen, or a consequence of polyclonal triggering, or both. The first alternative has not been excluded. In vivo triggering of B-lymphocytes in MS patients, by as yet not defined polyclonal activators, might well explain the production of the IgG antibodies against different viruses and bacteria as well as of autoantibodies demonstrable in MS. The proportion of these IgG antibodies in relation to the whole amount of intrathecally produced IgG is not known. A possible T-cell defect with loss of negative immunoregulatory function in the form of deficiency of suppressor T-cells leading to a secondary B-lymphocyte hyperactivity has been in focus since the demonstration of reduction in number and function of suppressor T-cells in MS (Antel et al., 1979, Reinherz et al., 1980).

The recent observation that the humoral immune response in MS is not restricted to IgG, warrants more careful evaluation of the B-lymphocyte response than mere quantitation of free Igs. Our results obtained with the indirect haemolytic PFC assay summarized in Table 6 have indeed demonstrated the occurrence in most MS patients of intrathecal IgM- and IgA-production, and have also shown that systemic B-lymphocyte hyperactivity is a common finding in this disease.

Table 6. B-cell Function in Multiple Sclerosis

Variable		Blood IgG	Blood IgA	Blood IgM	CSF IgG	CSF IgA	CSF IgM
Conc. of free Ig		Normal	Normal	Normal	↑↑	↑	↑↑
Ig-producing cells		↑	↑	↑	↑↑	↑	↑
Ig conc. in lymphocyte culture supernatants	–	↑	Normal	Normal	↑↑	↓	↓
PWM	+	↑	Normal	↑↑	↓↓	↓↓	↓↓

Analysis of IgG, IgA and IgM concentrations in supernatants from unstimulated lymphocyte cultures has confirmed that MS CSF-L have the capacity to produce Igs of these three classes, and IgG at a 3-fold higher rate than the corresponding PBL. However, MS CSF-L lacked the capacity to respond to PWm with increased Ig production, in contrast to the corresponding PBL. One explanation of the unresponsiveness of MS CSF-L might be that they are already maximally activated.

Future B-lymphocyte studies in MS employing inter alia similar techniques should concentrate on subclass definition of intrathecally produced IgG and of IgG antibodies, careful evaluation of the intrathecal IgM response with regard to antibody specificities, and characterization of antibodies in supernatants of CSF-L and corresponding PBL.

REFERENCES

Antel, J. P., Arnason, B. G. W., and Medof, M. E., 1979, Suppressor cell function in multiple sclerosis: correlation with clinical disease activity, Ann.Neurol., 5:338.

Brooks, B. R., Hirsch, R. L., and Coyle, P. K., 1983, Cellular and humoral immune responses in human cerebrospinal fluid, in: "Neurobiology of Cerebrospinal Fluid," Vol. 2, J. H. Wood, ed., Plenum Press, New York and London, pp. 263-329.

Cuzner, M. L., Davison, A. N., and Rudge, P., 1979, Preoteolytic enzyme activity of blood leukocytes and cerebrospinal fluid in multiple sclerosis, Ann.Neurol., 4:337.

Forsberg, P., and Kam-Hansen, S., 1983, Immunoglobulin-producing cells in blood and cerebrospinal fluid during the curse of aseptic meningoencephalitis, Scand.J.Immunol., 17:531.

Forsberg, P., henriksson, A., Link, H., and Ohman, S., 1983, Reference values for CSF-IgM, CSF-IgM/P-IgM ratio and IgM index, and its application on patients with multiple sclerosis and aseptic meningoencephalitis, Scand.J.Clin.Lab.Invest., (in press).

Frydén, A., Link, H., and Norrby, E., 1978, Cerebrospinal fluid and serum immunoglobulins and antibody titres in mumps meningitis and aseptic meningitis of other etiology, Infect.Immun., 21:851.

Goust, J.-M-, Hogan, E. L., and Arnaud, P., 1982, Abnormal regulation of IgG production in multiple sclerosis, Neurology, 32:228.

Henriksson, A., Kam-Hansen, S., and Andersson, R., 1981, Immunoglobulin-producing cells in CSF and blood from patients with multiple sclerosis and other inflammatory neurological diseases enumerated by protein-A plaque assay. J.Neuro immuno., 1:299.

Hijmans, W., and Schit, H. R. E., 1972, Immunofluorescene studies on immunoglobulins in the lymphoid cells of human peripheral blood, Clin.Exp.Immunol., 11:483.

Hood, L. E., Weissman, J., and Wood, W. B., 1978, "Immunology," The Benjamin/Cummings Publishing Co., Menlo Park, Ca., USA.

Kabat, E. A., Glusman, M., and Knaub, V., 1948, Quantitative estimation of the albumin and y-globulin in normal and pathological cerebrospinal fluid by immunochemical methods, Amer,J. Med., 4:653.

Kam-Hansen, S., Frydén, A., and Link, H., 1978, B and T cells in cerebrospinal fluid and blood in multiple sclerosis and acute mumps meningitis, Acta Neurol.Scand., 58:95.

Kam-Hansen, S., Link, H., Frydén, A., and Möller, E., 1979, Reduced in vitro response of CSF lymphocytes to mitogen stimulation in multiple sclerosis, Scand.J.Immunol., 10:161.

Kam-Hansen, S., 1980, Distribution and function of lymphocytes from the cerebrospinal fluid and blood in patients with multiple sclerosis, Acta Neurol.Scand., 62, Suppl. 62:1-81.

Laterre, E. C., Callewaert, A., Heremans, J. F., and Sfaello, Z., 1970, Electrophoretic morphology of gamma globulins in cerebrospinal fluid of multiple sclerosis and other diseases of the nervous system, Neurology, 20:982.

Laurenzi, M. A., Mavra, M., Kam-Hansen, S., and Link, H., 1980, Oligoclonal IgG and free light chains in multiple sclerosis demonstrated by thin-layer polyacrylamide gel isoelectric focusing and immunofixation, Ann.Neurol., 8:241.

Link, H., and Zettervall, O., 1970, Multiple sclerosis: Disturbed kappa: lambda chain ratio of immunoglobulin G in cerebrospinal fluid, Clin.Exp.Immunol., 6:435.

Link, H., and Müller, R., 1971, Immunoglobulins in multiple sclerosis and infections of the nervous system, Arch.Neurol., 25:326.

Mehta, P. D., Mehta, S. P., and Patrick, B. A., 1982, Identification of light chain type bands in CSF and serum oligoclonal IgG from patients with multiple sclerosis, J.Neuroimmunol., 2:119.

Norrby, E., 1978, Viral antibodies in multiple sclerosis, Progr.Med.Virol., 24:1.

Olsson, T., Kostulas, V., and Link, H., 1983, Demonstration of oligoclonal IgG bands in unconcentrated CSF by agarose isoelectric focusing and antiserum-avidin-biotin-peroxidase labelling, J.Neurol.Neurosurg.Psych., submitted.

Palmer, D. L., and Minard, B. J., 1976, IgG subgroups in cerebrospinal fluid in multiple sclerosis, N.Engl.J.Med., 294:447.

Reinherz, E. L., Weiner, H. L., Hauser, S. L., Cohen, S. A. Distaso, J. A., and Schlossman, S. F., 1980, Loss of suppressor T cells in active multiple sclerosis. Analysis with monoclonal antibodies, N.Engl.J.Med., 303:125.

Rinne, U. K., and Riekkinen, P. J., 1968, Esterase, peptidase and proteinase activities of human cerebrospinal fluid in multiple sclerosis, Acta.Neurol.Scand., 44:156.

Roström, B., 1981, Specificity of antibodies in oligoclonal bands in patients with multiple sclerosis and cerebrovascular disease, Acta Neurol.Scand., 63, Suppl.86:1-84.

Salier, J. P., Goust, J. M., Pandey, J. P., and Fudenberg, H. H., 1981, Preferential synthesis of the Glm(1) allotype of IgGl in the central nervous system of multiple sclerosis patients, Science., 213:1400.

Salier, J. P., Goust, J. M., Link, H., Pandey, J. P., Daveau, Maryvonne, and Fudenberg, H. H., 1983, Latent immunoglobulin G (Gm) allotypes: occurrence in the cerebrospinal fluid in some neuropathological states, J. Immunogenetics., 10:311.

Salmi, A., Reunanen, M., Ilonen, J., and Panelius, M., 1983, Intrathecal antibody synthesis to virus antigens in multiple sclerosis, Clin.Exp.Immunol., 52:241.

Sandberg-Wollheim, M., 1974, Immunoglobulin synthesis in vitro by cerebrospinal fluid cells in patients with multiple sclerosis, Scand.J.Immunol., 3:717.

Sindic, C. J. M., Cambiaso, C. L., Depré, A., Laterre, E. C., and Masson, P. L., 1982, The concentrations of IgM in the cerebrospinal fluid of neurological patients, J.Neurol.Sci., 55:339.

Strandberg-Pedersen, N., Kam-Hansen, S., Link, H., and Mavra, M., 1982, Specificity of immunoglobulins synthesized within the central nervous system in neurosyphilis, Acta.Path.Microbiol. Immunol.Scand.Sect.C., 90:97.

Tibbling, G., Link, H., and Ohman, S., 1977, Principles of albumin and IgG analysis in neurological disorders. I. Establishment of reference values, Scand.J.Clin.Lab.Invest., 37:385.

Trotter, J. L., and Brooks, B. J., 1980, Pathology of cerebrospinal fluid immunoglobulins, in: "Neurobiology of Cerebrospinal Fluid," Vol.1. J. H. Wood, ed., Plenum Press, New York and London, pp.465-486.

Walsh, M. J., and Tourtellotte, W. W., 1983, The cerebrospinal fluid in multiple sclerosis, in: "Multiple Sclerosis," J. F. Hallpike, C. W. M. Adams, and W. W. Tourtellotte, eds., Chapman and Hall, London, pp. 275-358.

Walsh, M. J., Tourtellotte, W. W., and Potvin, A. R., 1983, Central nervous system immunoglobulin synthesis in neurological disease, in: "Neurobiology of Cerebrospinal Fluid," Vol.2 J. H. Wood, ed., Plenum Press, New York and London, pp. 331-368.

Vartdal, F., Vandvik, B., and Norrby, E., 1980, Viral and bacterial antibody responses in multiple sclerosis, Ann.Neurol., 8:248.

VARICELLA-ZOSTER VIRUS LATENCY IN THE NERVOUS SYSTEM

Donald H. Gilden, Abbas Vafai, Yehuda Shtram,
Yechiel Becker, Mary Devlin and Mary Wellish

The Multiple Sclerosis Research Center of the Wistar Institute of Anatomy and Biology; the Departments of Neurology and Microbiology, University of Pennsylvania School of Medicine, Philadelphia, Pennsylvania; and the Department of Molecular Virology, The Hebrew University Jerusalem, Israel

INTRODUCTION

The cause of multiple sclerosis (MS) is not known. The main hypotheses are that the disease is produced by a virus or is a virus-triggered immunopathology. Although evidence to support the theory that MS is caused by a virus is only presumptive, there are four compelling reasons to search for virus in MS tissue. First, although MS does not occur until adult life, epidemiologic studies suggest that the causative agent is acquired in childhood.[1] An exposure period before or at the time of puberty followed by a long latent period before the onset of the disease is known to occur with other types of infection that lead to clinical CNS disease.[2] Furthermore, a point-source epidemic of MS occurred on the Faroe Islands from 1943-1960 during the British occupation;[3] thus MS on the Faroes would appear to be transmissible and probably infectious. Secondly, the cerebrospinal fluid (CSF) of patients with MS contains oligoclonal bands and increased levels of immunoglobulin (Ig)G irrespective of the amount of total protein. Only a few other diseases in humans are associated with oligoclonal bands and increased CSF-IgG; many are inflammatory and have also been shown to be infectious.[4] Third, progressive multifocal leucoencephalopathy is a human demyelinating disease in which a papovavirus produces a lytic infection of oligodendrocytes.[5] Fourth, numerous DNA and RNA viruses have been shown to be capable of producing demyelinating disease in rodents after experimental inoculation.[6]

Despite these findings, virus has not been detected in MS brain. There are no inclusion bodies nor have virus particles been detected by electron microscopy. Moreover, our examination of more than 20 MS brains by indirect immunofluorescence with multiple antiviral antisera has not revealed viral antigen. Growth of explanted MS brain cells in culture has not resulted in a viral cytopathic effect. Lysolecithin and polyethylene glycol-induced fusion of MS brain cells with a variety of indicator cells has also repeatedly failed to produce virus. Finally, analysis of MS brain cells experimentally infected with vesicular stomatitis virus did not result in the production of a virus pseudotype. Production of a non-neutralizing virus fraction would have suggested that a latent virus closely associated with cell membrane was present in the MS brain cells.

Because the classic methods to detect, isolate or rescue virus from MS tissue have failed, investigators are beginning to use the sensitive technique of nucleic acid hybridization to detect virus-specific gene sequences in tissue. Single-stranded nucleic acid "probes" that have been radioactively labelled to high specific activity are reacted in solution or with tissue (in situ hybridization) or in Southern blotting to detect complementary viral strands that will form double-stranded polynucleotides. To obtain probes that are "pure" (free of contaminating cellular or other DNA), the DNA fragments are cloned in plasmid or bacteriophage vectors.

This technique may have application in the search for virus genetic material in MS tissue. Probes prepared from a known virus or viral sequences van be prepared for hybridization with nucleic acid extracted from MS brain. The choice of virus probe should initially include each of those agents associated with the known human demyelinating disease, acute disseminated encephalomyelitis (ADEM). Thus measles, smallpx, rabies, influenza, mumps, varicellazoster virus (VZV), herpes simplex virus (HSV) and rubella are all candidates.

We have used nucleic acid hybridization techniques to detect VZV in human nervous system tissue. This virus is the causative agent of chickenpoz and shingles.[7] It is ubiquitious, acquired early in life and presumably becomes latest in dorsal root ganglia.[8] It is one of the most common causes of ADEM, particularly in the limited form of "cerebellar encephalitis" of childhood. Unless an individual has chickenpox or shingles just before death, it has not been possible to detect VZV in human ganglia. Tissue culture explant methods that have led to the successful isolation of HSV from human ganglia have not aided in the isolation of VZV from human dorsal root ganglia.[9]

The studies described below confirm for the first time that VZV is indeed latent in normal human dorsal root ganglia. Thus, there is a naturally occurring model system in humans with which to further study the biology of virus latency in the nervous system. As these methodologies develop, probes for VZV genetic material in MS tissue will also be made.

METHODS

Human Tissue, Virus and Cells

The trigeminal ganglion was obtained from a 53-year-old male who drowned. The sacral ganglion was removed from an 82-year-old male who sustained myocardial infarction and developed sacral distribution zoster immediately before death; VZV was isolated from his spinal fluid. Tissue was obtained 3 to 18 hours after death. After nerve roots were trimmed from the ganglia, they were washed three times with Dulbecco's minimum essential medium and stored at -70°C until DNA extraction.

The Web A (RIT) strain of VZV, originally isolated from the vesicle of a patient with chickenpox, has been propagated in our laboratory since 1977.[10] VZV was serially passaged in the BSC-1 line of African green monkey kidney cells by transferring infected cells into uninfected cultures at a ratio of 1:4.

Preparation of DNA from Tissue, from Uninfected BSC-1 Cells and from VZ Virions

The sacral ganglion was thoroughly Dounce-homogenized in 0.1 M NaCl, 50mM EDTA and 10 mM Tris-HCl, pH 7.5 at 4°C. The homogenized tissue was made to contain 0.5% sodium dodecyl sulfate (SDS) and 0.5 mg/ml pronase (Calbiochem-Behring) and incubated for 3 hours at 37°C with occasional agitation. The DNA was then phenol chloroform-extracted, ethanol-precipitated and resuspended in TE buffer (10 mM Tris-HCl, pH 7.5, 1 mM EDTA). The trigeminal ganglion was partially Dounce-homogenized at 4°C and then digested with SDS-pronase for 18 to 24 hours at 37°C. The tissue was then rehomogenized, incubated for 3 hours at 37°C with an additional 250 μg/ml of pronase, and the DNA was phenol chloroform-extracted, ethanol-precipitated and dissolved in TE buffer. Uninfected BSC-1 cells were washed twice with a solution of 0.2 M Tris-HCl, pH 7.5, 0.15 M NaCl, scraped into TE buffer, and incubated for 3 hours at 37°C with SDS-pronase, phenol chloroform-extracted, ethanol-precipitated and resuspended in TE buffer. VZ virions were prepared from infected BSC-1 cells according to the method of Wroblewska et al.[11] To extract viral DNA, the virions were incubated for 3 hours at 37°C with SDS-pronase. Viral DNA was phenol chloroform-extracted, ethanol-precipitated and resuspended in TE buffer. All extracted DNA, except that from virions, was then incubated with pancreatic ribonuclease A (50 μg/ml) for 2 hours at 37°C, phenol chloroform-extracted, ethanol-precipitated and dissolved in TE buffer. The concentration of the DNA was determined by measuring the absorbance at 260 nm (1 µg equal 0.02 A_{260}). Approximately 300 µg of DNA was obtained from the trigeminal ganglion, and 140 µg from the necrotic inflamed sacral ganglion.

Preparation and Cloning of VZV-DNA

To clone VZV-DNA, viral DNA was extracted from VZV-infected BSC-1 cells with 0.25% Triton X-100-0.2 M NaCl, and purified by isopycnic centrifugation in CsCl as described by Gilden et al.[12] Viral DNA was cloned in the plasmid pBR322 at the BamHI site, and the plasmid was amplified and purified according to the methods described by Maniatis et al.[13]

Nick-translation of Cloned VZV-DNA

The cloned VZV-DNA was labelled with α-^{32}P dATP and α-^{32}P dCTP (Amersham; 3000 Ci/mM) according to the nick-translation procedures of Rigby[14] and Maniatis.[15] The reaction mixture (50 µl) contained 0.125 Ci of each labelled triphosphate, 200 ng DNA, nick-translation buffer (50 mM Tris-HCl, pH 7.8 5 mM $MgCl_2$, 50 µg/ml bovine serum albumin (BSA), 0.01 mM DL-dithiothreitol), 500 pg DNase (Worthington), 1 mM dGTP, 1 mM dTTP, and 5 units of E. coli DNA polymerase I (Boehringer-Mannheim). Incubation was at 15°C for 90 minutes. The reaction was terminated by the addition of 200 µl of 0.1 M NaCl, 10 mM Tris-HCl, pH 7.4, 1 mM EDTA, 0.1% SDS containing 100 µg of sonicated salmon sperm DNA. The solution was passed through a Sephadex G-50 column, and the peak fractions containing the radio-labelled DNA were pooled, ethanol-precipitated, centrifuged and dissolved in 600 to 800 µl of TE buffer. ^{32}P-labelled DNA probes with specific activities of 6-8 x 10^8 cpm/µg of DNA were used for DNA-DNA hybridization.

Preparation of Blots

BamHI-cleaved DNA fragments were separated and transferred to nitrocellulose filters according to the methods described by Southern[16] and Wahl.[17] Twenty to 40 units of the restriction endonuclease BamHI were used to digest 10 to 20 µg of cell DNA in 50 mM NaCl, 7 mM Tris-HCl, pH 7.5, 2 mM $MgCl_2$ and 2 mM 2-mercapto-ethanol. After incubation for 2 hours at 37°C, an additional 10 to 20 units of enzyme were added and digestion was continued for 1 hour. The reaction was stopped by the addition of 5 µl of a solution containing bromophenol blue (0.005%), 30% sucrose and 50 mM EDTA. DNA fragments were separated in 0.6% agarose gels in Loening buffer (39 mM Tris-base, 34 mM sodium acetate, 2 mM EDTA, 0.18% acetic acid). After electrophoresis, gels were stained with ethidium bromide. Gels were then blotted onto nitrocellulose paper according to the procedures described by Southern[16] and Wahl.[17]

Hybridization

The hybridization technique was essentially that of Wahl et al.[17] Nitrocellulose filters were pretreated at 50°C with 15 to 20 ml of 20% formamide, 0.1% SDS, 4X SSC (0.6 M NaCl, 60 mM sodium citrate), 5X Denhardt's solution (0.1% Ficoll, 0.1% polyvinylpyrolidine, 0.1% BSA), and 1 mM EDTA containing 250 µg/ml sonicated denatured salmon sperm DNA for approximately 18 hours. Excess prehybridization solution was removed and hybridization was conducted at 50°C for 42 hours. The hydridization solution consisted of 42% formamide, 10% dextran sulfate, 1X Denhardt's solution, 0.1% SDS, 5X SSC and 1 mM EDTA containing 100 µg/ml denatured sonicated salmon sperm DNA and heat-denatured probe (1.2-1.6 x 10^8 cpm per filter). The filters were washed for 5 minutes at room temperature with 0.1X SSC and 0.1% SDS followed by four 15-minute washes with the same solution at 50°C and two 15-minute washes with 2X SSC and 0.1% SDS at 65°C. Filters were dried, wrapped in Saran Wrap and autoradiographed using Kodak XAR-5 film in cassettes containing Dupont Cronex intensifying screens at -70°C for 3 hours to 7 days.

RESULTS

A BamHI-cleaved VZV-DNA fragment was cloned into the plasmid vector pBR322. The size of the VZV-DNA fragment is 9.7 kb, which corresponds to the C fragment produced by BamHI digestion of VZ virion DNA (Fig. 1) and maps at 0.16-0.25 map units.[18] The cloned VZV-DNA fragment was labelled with ^{32}P (4-8 x 10^8 cpm/µg DNA) by nick-translation and hybridized to DNAs extracted from a single sacral ganglion removed from an 82-year-old male who had sacral distribution zoster immediately before death and from a normal trigeminal ganglion of a 53-year-old male who drowned.

Hybridization of the VZV-DNA fragment C to DNAs from sacral (lane 3) and trigeminal ganglia (lane 4) was observed. However, no hybridization was detected when VZV-DNA fragment C was hybridized to cell DNA (lane 5) or to DNA extracted from herpes simplex virus (HSV), cytomegalovirus (CMV) and Epstein-Barr virus (EBV) (not shown), thus indicating the presence of specific VZV-DNA sequences in human sensory ganglia. Nick-translated ^{32}P-labelled pBR322 also did not hybridize with DNA from the human ganglia.

DISCUSSION

We have used DNA-DNA hybridization to demonstrate the presence of VZV-DNA sequences in human sensory ganglia. The detection of VZV-DNA sequences in the sacral ganglion of a man with sacral distribution zoster immediately before death not only supports previous studies showing VZV antigen[19] and the isolation of VZV[20] from

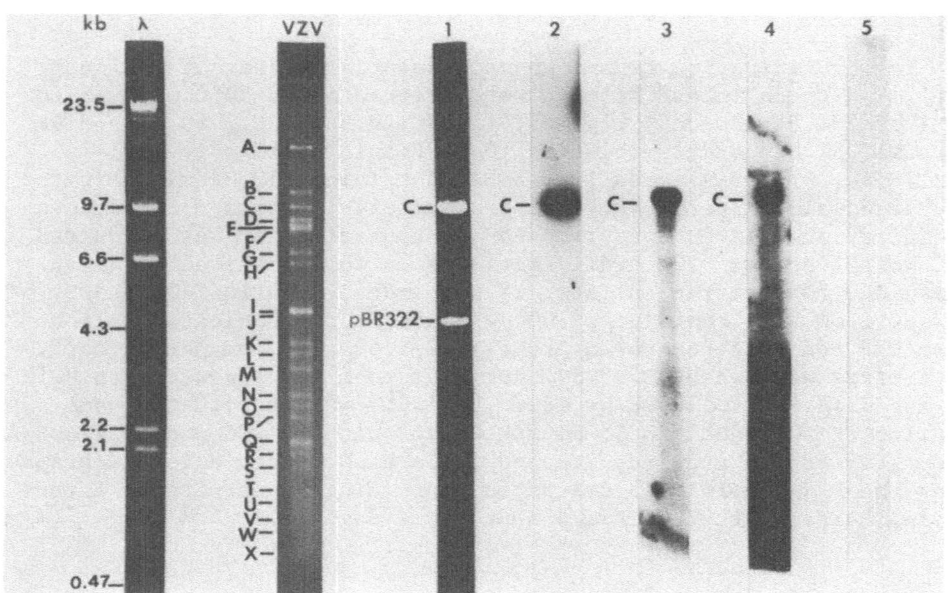

Fig. 1. Detection of varicella-zoster virus (VZV) DNA sequences in human sensory ganglia. VZV-DNA was cleaved with the BamHI restriction endonuclease and DNA fragments were cloned in the plasmid vector pBR322 as described in Methods. VZV-DNA cloned fragment C (lane 1) was amplified in E. coli HB101 and purified as described in Methods. Viral DNA fragment C was then labelled with ^{32}P and hybridized to BamHI-cleaved VZV-DNA (lane 2), sacral ganglion DNA (lane 3), trigeminal ganglion DNA (lane 4), and monkey kidney cell DNA (lane 5). Left lane shows the bacteriophage λ Hind III-cleaved DNA fragments used as a molecular size marker. The VZV lane shows the pattern of DNA fragments produced by BamHI restriction endonuclease. The bands marked C in lanes 3 and 4 represent the presence of VZV-DNA sequences in human sensory ganglia.

acutely infected human ganglia, but also extends the techniques available for detection of VZV in human tissue to the nucleic acid level. A common site for zoster development is on the face, and thus the demonstration of VZV genetic material in a trigeminal ganglion is not surprising. More ganglia obtained from all levels of the human neuraxis, including autonomic nervous system ganglia, must be tested to determine the prevalence of VZV-DNA in ganglia.

The studies described here indicate that pure well-characterized probes prepared by recombinant DNA techniques can be used to identify specific viral DNA sequences in normal nervous system tissue when the classic virologic methods of co-cultivation and cell fusion have failed. As more VZV-specific fragments become available, detection of viral sequences in the nervous system tissue will be enhanced. These probes will then be applied not only to further study the latent state of VZV in human ganglia, but also will be used to search for VZV sequences in MS tissue.

Acknowledgements

We thank Kathleen Reilly, Helene van Trieste and Dr. Charles Reinhold for assistance in obtaining human ganglia, Gregory Dressler for expert technical help, Audrey Gilden for printing the photograph, Dr. William Stroop and Marina Hoffman for a critical review and Suzanne Amrhein for her help in the preparation of this manuscript. This work was supported by grants NS-10036 from the National Institutes of Health and 894-D4 from the National Multiple Sclerosis Society.

REFERENCES

1. M. Alter, U. Leibowitz, and J. Speer, Risk of multiple sclerosis related to age at immigration to Israel, Arch.Neurol., 15:234-237 (1966).
2. D. H. Gilden, Slow virus diseases of the CNS, Postgrad.Med., 73:99-118 (1983).
3. J. F. Kurtzke and K. Hyllested, Multiple sclerosis in the Faroe Islands: I. Clinical and epidemiological features, Ann. Neurol., 5:6-21 (1978).
4. J. R. Miller, A. M. Burke and C. T. Bever, Occurrence of oligoclonal bands in multiple sclerosis and other CNS diseases, Ann.Neurol., 13:53-58 (1982).
5. L. P. Weiner, R. T. Johnson and R. M. Herndon, Viral infections and demyelinating diseases, New Eng.J.Med., 288:1103-1110 (1973).
6. M. C. Dal Canto and S. G. Rabinowitz, Experimental models of virus-induced demyelination of the central nervous system, (Ann.Neurol., 11:109-127 (1981).
7. B. D. Davis, R. Dulbecco, H. N. Eisen and H. S. Ginsberg, in: "Microbiology," Harper and Row, eds., Hagerstown, Maryland, 1068 (1980).
8. R. E. Hope-Simpson, The nature of herpes zoster: a long-term study and a new hypothesis, Proc.Roy.Soc.Med., 58:9-20.
9. S. A. Plotkin, S. Stein, M. Snyder and R. Immesoete, Attempts to recover varicella virus from ganglia, Ann.Neurol., 2:249 (1977).
10. D. H. Gilden, Z. Wroblewska, V. Kindt, K. G. Warren and J. S. Wolinsky, Varicella-zoster virus infection of human brain cells and ganglion cells in tissue culture, Arch.Virol., 56:105-117 (1978).

11. Z. Wroblewska, M. Devlin, K. Reilly, H. van Trieste, M. Wellish, and D. H. Gilden, The production of varicella zoster virus antiserum in laboratory animals, Arch.Virol., 74:233-238 (1982).
12. D. H. Gilden, Y. Shtram, A. Friedmann, M. Wellish, M. Devlin, A. Cohen, M. Fraser, and Y. Becker, Extraction of cell-associated varicella zoster virus with Triton X-100-NaCl, J.Virol.Meth., 4:263-276 (1982).
13. T. Maniatis, E. F. Fritsch and J. Sambrook, "Molecular cloning, A laboratory manual", Cold Spring Harbor Laboratory, 77-96 (1982).
14. P. W. J. Rigby, M. Dieckmann, C. Rhodes and P. Berg, Labelling deoxyribonucleic acid to high specific activity in vitro by nick translation with DNA polymerase I., J.Mol.Biol., 113:237-251 (1977).
15. T. Maniatis, A. Jeffrey and D. G. Kleid, Nucleotide sequence of the rightward operator of phage, Proc.Natl.Acad.Sci., 72:1184-1188 (1975).
16. E. M. Southern, Detection of specific sequences among DNA fragments separated by gel electrophoresis, J.Mol.Biol., 98:503-517 (1975).
17. G. M. Wahl, M. Stern and G. R. Stark, Efficient transfer of large DNA fragments from agarose gel to diazobenzyloxymethyl-paper and rapid hybridization by using dextran sulfate, Proc.Natl.Acad.Sci., 76:3683-3687 (1979).
18. J. R. Ecker and R. W. Hyman, Varicella zoster virus DNA exists as two isomers, Proc.Natl.Acad.Sci., 79:156-160 (1982).
19. M. M. Esiri and A. H. Tomlinson, Herpes zoster. Demonstration of virus in trigeminal nerve and ganglion by immunofluorescence and electron microscopy, J.Neurol.Sci., 15:35-48 (1972).
20. F. O. Bastian, A. S. Rabson, C. L. Yee and T. S. Tralka, Herpes-virus varicellae. Isolated from human dorsal ganglia, Arch.Pathol., 97:331-332 (1974).

DISCUSSION

Above all I would like to congratulate Dr. Gilden on his successful approach to viral pathology of the central nervous system (CNS) by application of the sophisticated nucleic acid hybridization technique. I feel that we can expect from this technique very important progress of knowledge in the near future.

With regard to the relevance of these results for etiopathogenesis of multiple sclerosis (MS), however, one has to say that the primary Varicella-Zoster-infections of CNS are very different in nature from MS. They are mostly monophasic diseases, only sometimes comparable to acute MS but never showing typical plaque-formation. They are also not linked to HLA-factors as MS. May I turn therefore with a few remarks to the problem of viral origin of MS and demyelination in general, as recently summarized by B. Waksman.

A viral infection could be responsible for MS in 3 ways:

1. Direct infection and damage of oligodendrocytes would produce myelin breakdown and probably secondary inflammation.
2. Antigens encoded by persistent virus genomic material might produce immunological response in the form of primary inflammation and secondary demyelination.
3. A virus might induce an autoimmune response to some components of the oligodendroglia on the myelin sheath.

An animal model corresponding to the first possibility would be JHM virus and visna. JHM mutants appear to infect oligodendrocytes and thus to produce demyelination with certain degree of inflammation. An animal model for the second possibility is provided by Theiler's virus in mice. In this model of persistence of viral genome there is continuous expression of viral antigens which is the prerequisite for production of an immunological disease. With regard to the third possibility several new observations indicate that viruses infecting the CNS can induce autoimmunity against CNS (myelin) antigens.

If we turn to MS one has to state that a unique virus producing demyelination has not been isolated from neural tissue of MS-patients. Apparent viral infections of oligodendrocytes as e.g. in PML are totally different in pathogenesis and pathomorphology. Several reports of putative causal agents had been published which all appear non-specific.

By sensitive variants of the nucleid acid hybridization technique recently two common viruses had been demonstrated in CNS tissue. Haase et al. found persistent measles virus genome in 6 or 12 MS-brains and 1 of 7 normal controls. Frazier et al. demonstrated herpes simplex genome in 4 of 5 MS and 3 of 6 normals. In some instances the complete viral genome was present, in others only part of the genome; both grey matter and white matter were involved. So far no expression of antigens of measles or herpes has been found in MS-brains. The fact that MS-patients present circulating antibodies against common viruses with frequently elevated titers appears to depend on the presence of HLA-A 3 or B 7-factors in the individual and not on the presence of MS. On the other hand cell-mediated immunity against measles virus appears to be low or absent in MS-patients. Responses of MS-patients to myelin or oligodendrocyte-antigens, in particular as an expression of the response to viral invasion of the CNS are not yet extensively studied. The introduction of T-lymphocyte cloning techniques will allow a new and perhaps successful attack on the assessment of specific reactivities to viral and neural antigens. De. Johnson recently in a study of measles cases of Lima has shown that myelin basic protein (MBP) appears early in the spinal fluid associated with pleocytosis as a direct evidence of tissue damage and that virus is present in the spinal fluid cells.

This may lead to rapid sensitization against neural antigens with lymphocytic reactivity in a high percentage of viral encephalitis cases. Therefore the hypothesis seems to be justified that autoimmunization as a consequence of early virus infections of the CNS in individuals with appropriate abnormalities of immune regulation may provide the basis of a chronic relapsing demyelinating disease.

THE MYELIN ASSOCIATED GLYCOPROTEIN IN DEMYELINATING DISEASES

Norman Latov and Eduardo Nobile-Orazio

Department of Neurology, Columbia University College of Physicians and Surgeons, 630 West 168th Street, New York 10032, U.S.A.

INTRODUCTION

The myelin associated glycoprotein (MAG) is an integral protein of both central nervous system (CNS) and peripheral nervous system (PNS) myelin. Its molecular weight is approximately 100K Daltons (Quarles, 1979; Figlewicz et al., 1981), and its concentration is 2.7μg/mg protein in adult rat brain and 0.65μg/mg in sciatic nerve (Johnson et al., 1982). MAG is approximately one-third carbohydrate by weight and contains fucose, mannose, galactose, N-acetylglucosamine and N-acetylneuraminic acid (Barbarash et al., 1981).

The localization of MAG, its structural arrangement within the myelin sheath, and its function are currently under investigation. Immunocytochemical studies with antibodies to MAG yield varying results and partly depend on the antiserum, tissue and fixative used. In the CNS, immunostaining of the periaxonal region (Sternberger et al., 1979) or of compact myelin (Webster et al., 1983) is reported, and in the PNS, immunostaining of periaxonal and paranodal regions and of the Schmidt Lantermann clefts and outer mesaxon is observed Trapp et al., 1982). The mouse monoclonal antibody GEN-S8 immunostains the periaxonal and outer regions of PNS myelin, whereas the human anti-MAG M-proteins and the mouse monoclonal antibody anti-HNK-1 (anti-Leu-7, Becton Dickenson) immunostain PNS myelin diffusely (Abrams et al., 1982; Stefansson et al., 1983; Takatsu et al., submitted for publication; Nobile-Orazio et al., submitted for publication). However, the M-proteins and anti-HNK-1 may react with carbohydrate determinants in MAG which are shared by other PNS myelin proteins and their immunostaining may not be specific for MAG (Nobile-Orazio et al., submitted for publication). It may be that different determinants of MAG are differently oriented or exposed in

CNS and PNS myelin and in different regions of the myelin sheath, or that MAG is preferentially concentrated in certain myelin regions. The function of MAG is unknown, but the antibodies that selectively immunostain the periaxonal and outer regions of myelin may be recognizing determinants which are bound and shielded in compacted myelin but exposed on the unopposed periaxonal and outer myelin regions. Alternatively, MAG may be involved in maintenance of the myelin-axcn junction (Sternberger et al., 1979; Trapp et al., 1982).

MAG in Demyelinating Diseases of the Central Nervous System

In an immunocytochemical study of multiple sclerosis, there was decreased immunostaining of MAG in white matter surrounding acute lesions. The same areas appeared normal when stained for myelin or when immunostained for basic protein (Itoyama et al., 1980). This pattern of immunostaining was found to be specific for multiple sclerosis and was not found in other demyelinating conditions including in progressive multifocal leukoencephalopathy (Itoyama et al., 1982), idiopathic polyneuritis (Schrober et al., 1981) and experimental allergic encephalomyelitis (Itoyama et al., 1982). These observations suggested that MAG may be affected early in demyelination and that it may be a primary target.

MAG in Demyelinating Peripheral Neuropathy

In some patients with peripheral neuropathy there are IgM monoclonal antibodies or M-proteins that react with MAG (Latov, 1983; Latov, et al., 1980; 1981; Braun et al., 1982; Mendell et al., 1982; Steck et al., 1983; Stefansson et al., 1983; Saito et al., 1983; Nobile-Orazio et al., 1983). Morphological studies of affected sural nerves show demyelination, widening of myelin lamellae, and deposition of M-protein on the myelin sheaths (Nemni et al., 1983; Steck et al., 1983; Stefansson et al., 1983; Takatsu et al., submitted for publication). Therapy directed at lowering the M-protein concentration may be beneficial (Latov et al., 1980). Studies using tissue and species specificity, peptide mapping, deglycosylation, and competitive binding assays, suggest that the M-proteins from different patients all react with the same determinant of MAG, and that the antigenic determinant contains carbohydrate residues (Shy et al., submitted for publication).

Patients with anti-MAG M-proteins suffer from neuropathy rather than from CNS disease, but in two of four patients tested there were abnormalities of the visual evoked responses suggestive of central demyelination. It is now known why the CNS is relatively spared but the concentration of the M-proteins in brain may be kept low by the blood-brain barrier, or in CNS myelin the reactive MAG determinant may be shielded within the myelin sheath and binding of the M-protein may be hindered.

In one patient studied, *in vitro* production of anti-MAG M-protein was stimulated by OKT4+ T-helper cells and suppressed by OKT8+ suppressor T-cells (Latov et al., in preparation). The system may be analagous to a carrier/hapten system where the antigenic determinant is the hapten and the protein the carrier (Herzenberg et al., 1983). However, it is not known whether the carrier required for T-cell recognition is the MAG protein or another protein in PNS myelin or elsewhere sharing the antigenic determinant. This may be important as in some systems, the anti-hapten antibody response may be suppressed by repeated immunizations with the carrier protein alone (Herzenberg et al., 1983).

In preliminary experiments, passive transfer of serum from a patient with anti-MAG M-proteins induced focal demyelination in recipient cat sciatic nerve (Hays et al., 1983), but the mechanism of action or the relevance to the human disease is unclear. The immunoregulatory mechanisms involved in elaboration of the M-protein, its role in the neuropathy, and the possible mechanism of action remain to be elucidated.

MAG and Natural Killer Cells

It was recently reported that the mouse monoclonal antibody anti-HNK-1 (anti-Leu-7, Becton Dickenson) which reacts with some human natural killer (NK) cells, also binds to MAG (Heberman et al., 1979; Abo and Balch, 1981; Helfand et al., 1983). This could be important as immune reactivity to NK cells could secondarily involve myelin and cause demyelination, or since MAG is an integral myelin protein which binds to other myelin components, NK could also bind to these and cause demyelination. Binding of NK cells to myelin could occur in the absence of evidence for immune reactivity as traditionally demonstrated by specific antibody or proliferative response, but could be dependent on factors such as accessibility or alterations of the myelin membrane. NK cells could potentially also have a protective effect as if infectious or immune agents were to bind to myelin *via* receptors for MAG, NK cells could bind to these and prevent demyelination.

In recent experiments, no reactivity was detected between human peripheral blood lymphocytes and a human anti-MAG M-protein or three mouse monoclonal anti-MAG antibodies (Latov et al., submitted for publication). Also, in patients with anti-MAG M-proteins, the number of circulating HNK-1+ cells was the same as in controls. These suggest that the determinant which binds to anti-HNK-1 is different than that recognized by the human M-proteins, and that MAG and NK cells may have only the HNK-1+ determinant in common.

There is no evidence to date that NK cells or HNK-1+ lymphocytes are important in demyelination. The number of HNK-1+ cells in patients with multiple sclerosis is the same as in normal subjects, and studies of NK cell function in patients have yielded varying results (Rice et al., 1983). However, lymphocytes cytotoxic to oligodendroglia, myelin, or to cells modified with myelin antigens have been described in multiple sclerosis (Berg and Kallen, 1964; Eggers et al., 1981; Frick, 1982; Halpern et al., 1969; Hauw et al., 1975), and NK cells may show considerable phenotypic and functional heterogeneity (Hercend et al., 1983). The possible role of HNK-1+ cells in experimental demyelination and in demyelinating diseases of the central and peripheral nervous system warrants further investigation.

REFERENCES

Abo, T., and Balch, C. M., 1981, A differentiation antigen of human NK and K cells identified by a monoclonal antibody (HNK-1), J.Immunol., 127:1024-1029.

Abrams, G. M., Latov, N., Hays, A.P., Sherman, W. H., and Zimmerman, E. A., 1982, Immunocytochemical studies of human peripheral nerve with serum from patients with polyneuropathy and paraproteinemia, Neurology(NY), 32:821-826.

Barbarash, G. R., Figlewicz, D. A., and Quarles, R. H., 1981, Myelin associated glycoprotein: purification and partial characterization (abstract), Am.Soc.Neurochem., 12:165.

Berg, O., and Kallen, B., 1964, Effect of mononuclear blood cells from multiple sclerosis patients on neuroglia in tissue culture, J.Neuropath.Exptl.Neurol., 23:550-559.

Braun, P. E., Frail, D. E., and Latov, N., 1982, Myelin associated glycoprotein is the antigen for a monoclonal IgM in polyneuropathy, J.Neurochem., 39:1261-65.

Eggers, A. E., Tarmin, L., Plank, C. R., and Gamboa, E. T., 1981, Hyperactivity to myelin basic protein in multiple sclerosis, J.Neurol.Sci., 52:385-390.

Figlewicz, D. A., Quarles, R. H., Johnson, D., Barbarash, G. R., and Sternberger, N. H., 1981, Biochemical demonstration of the myelin associated glycoprotein in the peripheral nervous system, J.Neurochem., 37:749-758.

Frick, E., 1982, Cell mediated cytotoxicity by peripheral blood lymphocytes against basic protein of myelin encephalitogenic peptide, cerebrosides, and gangliosides in multiple sclerosis, J.Neurol.Sci., 57:55-66.

Halpern, B., Bakouche, P., and Martial-Lasfargues, C., 1969, Destruction des cellules nerveuses, cultivees in vitro par les lymphocytes de malades atteints de sclerose en plaques, Press.Med., 77:2103-2106.

Hauw, J. J., Berger B., and Escourolle, R., 1975, Etude de la cytotoxicite lymphocytaire sanguine au cours de la sclerose en plaques, in: "Immunopathologie du Systeme Nerveus", M. E. Shuller, ed., Inserm, Paris.

Hays, A. P., Takatsu, M., Latov, N., and Sherman, W. H., 1983, Focal demyelination of cat sciatic nerve induced by intraneural injection of serum from patients with polyneuropathy and monoclonal IgM reactive with myelin associated glycoprotein, (abstract), J.Neuropath.Exptl.Neurol., 42:349.

Helfand, S. T., McCarry, R., Eaton, L., and Roder, J. C., 1983, Human natural killer cells and myelin share a common antigen, (abstract), Neurology(NY), 33(suppl.2):107.

Herberman, R. B., DJeu, J. Y., Kay, D., Ortaldo, J. R., Riccardi, C., Bonnard, G. D., Holden, H. T., Fagnani, R., Santoni, A., and Puccetti, P., 1979, Natural killer cells: characteristics and regulation of activity, Immunol.Rev., 44:43-70.

Hercend, T. Reinherz, E. L., Meuer, S., Schlossman, S. F., and Ritz, J., 1983, Phenotypic and functional heterogeneity of human cloned natural killer cell lines, Nature, 301:158-160.

Herzenberg, L. A., Tokuhisa, T., and Hayakawa, K., 1983, Epitope specific regulation, in: "Annual Review of Immunology, Vol.1", C. Garrison and H. Metzger, eds., Annual Review Inc., Ca.

Itoyama, Y., Sternberger, N. H., Webster, H. de F., Quarles, R. H., Cohen, S. R., and Richardson, Jr., E. P., 1980, Immunocytochemical observations on the distribution of myelin-associated glycoprotein and myelin basic protein in multiple sclerosis lesions, Ann.Neurol., 7:167-177.

Itoyama, Y., and Webster, H. de F., 1982, Immunocytochemical study of myelin associated glycoprotein and basic protein in acute experimental allergic encephalomyelitis, J.Immunol., 3:351-364.

Itoyama, Y., Webster, H. de F., Sternberger, N. H., Richardson, E. P., Walker, D. L., Quarles, R. H., and Padgett, B. L., 1982, Distribution of papovavirus, myelin associated glycoprotein, and myelin basic protein in progressive multifocal leukoencephalopathy lesions, Ann.Neurol., 11:396-407.

Johnson, D., Quarles, R. H., and Brady, R. O., 1982, A radioimmunoassay for the myelin associated glycoprotein, J.Neurochem., 39:1356-62.

Latov, N., 1983, Immunological abnormalities associated with chronic peripheral neuropathies: plasma cell dyscrasia and neuropathy, in: "First International Congress on Neuroimmunology", F. Spreafico and P. O. Behan, eds., Raven Press, N.Y. (in press).

Latov, N., Braun, P. E., Gross, R. B., Sherman, W. H., and Chess, L., 1981, Plasma cell dyscrasia and peripheral neuropathy; identification of the myelin antigens that react with human paraproteins, Proc.Natl.Acad.Sci.(U.S.A.), 78:7139-7142.

Latov, N., Godfrey, M., Nobile-Orazio, E., Messito, M. J., Perman, G., Abrams, J., Freddo, L., and Chess, L., Shared antigenicity between natural killer cells and the myelin associated glycoprotein, (submitted for publication).

Latov, N., Godfrey, M., Thomas, Y., Nobile-Orazio, E., Freddo, L., Perman, G., Abrams, J., and Chess, L., In vitro production of anti-MAG M-proteins; regulation by T-cells, (in preparation).

Latov, N., Sherman, W. H., Nemni, R., Galassi, G., Shyong, J. S., Penn, A. S., Chess, L., Olarte, M. R., Rowland, L. P., and Osserman, E. F., 1980, Plasma cell dyscrasia and peripheral neuropathy with a monoclonal antibody to peripheral nerve myelin, N.Engl.J.Med., 303:618-621.

Mendell, J. R., Sahenk, Z., Whitaker, J. N., Pittman, G., 1982, Monoclonal IgM and peripheral neuropathy: studies on passive transfer and antibody characterization, (abstract), Neurology (N.Y.), 32(2):A183.

Nemni, R., Galassi, G., Latov, N., Sherman, W. H., Olarte, M. R., and Hays, A. P., 1983, Polyneuropathy in nonmalignant IgM plasma cell dyscrasia: a morphological study, Ann.Neurol., 14:43-54.

Nobile-Orazio, E., Hays, A. P., Latov, N., Perman, G., Golier, J., Shy, M., and Freddo, L., Reactivity of mouse and human monoclonal anti-MAG antibodies; antigenic specificity and immunofluorescence studies, (submitted for publication).

Nobile-Orazio, E., Vietorisz, T., Messito, M. J., Sherman, W. H., and Latov, N., 1983, Anti-MAG IgM antibodies in patients with neuropathy and IgM M-proteins. Detection by ELISA, Neurology (Cleveland), 33:939-942.

Quarles, R. H., 1979, Glycoproteins in myelin and myelin related membranes, in: "Complex Carbohydrates of Nervous Tissue", R. V. Margolis and R. K. Margolis, eds., Plenum Press, New York.

Rice, G. P. A., Casali, P., Merigan, T. C., and Oldstone, M. B. A., 1983, Natural killer cell activity in patients with multiple sclerosis given interferon, Ann.Neurol., 14:333-338.

Saito, T., Sherman, W. H., and Latov, N., 1983, Specificity and idiotype of M-proteins that react with MAG, J.Immunol., 130:2496-99.

Schrober, R., Itoyama, Y., Sternberger, N. H., Trapp, B. D., Richardson, E. P., Asbury, A. K., Quarles, R. H., and Webster, H. de F., 1981, Immunocytochemical study of Po glycoprotein, P1 and P2 basic proteins, and myelin associated glycoprotein in lesions of idiopathic polyneuritis, Neuropath.Applied Neurobiol., 7:421-34.

Shy, M., Vietorisz, T., Nobile-Orazio, E., Freddo, L., and Latov, N., Specificity of human anti-MAG M-proteins; peptide mapping, deglycosylation, and competitive binding studies, (in preparation).

Steck, A. J., Murray, N., Meier, C., Page N., and Perruisseau, G., 1983, Demyelinating neuropathy and monoclonal IgM antibody to myelin associated glycoprotein, Neurology(NY), 33:19-23.

Steffanson, K., Marton, L., Antel, J. P., Wollmann, R. L., Roos, R. P., Chejfec, G., and Arnason, B. G. W., 1983, Neuropathy accompanying IgM monoclonal gammopathy, Acta Neuropath. (Berlin), 59:255-261.

Sternberger, N. H., Quarles, R. H., Itoyama, Y., Webster, H. de F., 1979, Myelin associated glycoprotein demonstrated immunocytochemically in myelin and myelin forming cells of developing rat, Proc.Natl.Acad.Sci.(USA), 76:1510-1514.

Takatsu, M., Hays, A. P., Latov, N., Abrams, G. M., Sherman, W. H., Nemni, R., Nobile-Orazio, E., Saito, T., and Freddo, L., Immunofluorescence study of patients with neuropathy and IgM M-proteins, (submitted for publication).

Trapp, B. D., and Quarles, R. H., 1982, Presence of the myelin associated glycoprotein correlates with alterations in the periodicity of peripheral myelin, J.Cell.Biol., 92:877-882.

Webster, H. de F., Palkovits, C. G., Stower, G. L., Farilla, J. T., Frail, D. E., and Braun, P. E., 1983, Myelin associated glycoprotein: electron microscopic immunocytochemical localization in compact developing and adult central nervous system myelin, J.Neurochem., 41:1469-1479.

DISCUSSION

Dr. Latov's contribution appears of great general relevance in immunopathological conditions. Recent evidence indicates a defect in immune regulation in MS patients. This mainly involves a defect of suppressor cells and of Natural Killer (NK) cells. An explanation of this phenomenon recently proposed by Arnason (Acta Neuropath. Suppl. IX) is the presence of shared antigenic determinants between T-cell sub-populations and myelin/oligodendroglia. In a recent study we were able to show that the monoclonal Leu 7 (HNK 1) antibody, a marker which recognizes NK-cells and a fraction of suppressor cells, also detects an antigen present in human central and peripheral myelin, in oligodendrocytes and in low concentration also in other components of the CNS (Schuller, Petrovic, Gebhart, Lassmann, Kraft and Rumpold; Nature: in press). Immunochemical evidence indicates, that this Leu 7 antibody recognizes an epitop of the Myelin Associated Glycoprotein (MAG) (Helfand et al.; Neurology 33. Suppl. II, p.107). Thus it appears reasonable that an immune reaction against this myelin protein may simultaneously induce a defect in suppressor and NK-cells and result in imbalance of immune regulation. This may be a very important aspect of MS pathogenesis.

CEREBROSPINAL FLUID IMMUNOELECTROPHORETIC FINDINGS IN MULTIPLE SCLEROSIS

O. J. Kolar, P. H. Rice, M. R. Farlow, and J. H. Wright

Department of Neurology, Multiple Sclerosis Research Laboratory, Indiana University School of Medicine and Computing Services, Indiana University-Purdue University at Indianapolis, U.S.A. 46223

INTRODUCTION

In 1962, Laterre et al.,[1] and later Dencker[2] reported the presence of a second precipitation arc, in a parallel position to the precipitate of immunoglobulin G (IgG), in cerebrospinal fluid (CSF) immunoelectrophoresis (IE) of some patients with multiple sclerosis (MS). This doubling of the CSF IgG precipitate was not demonstrated in the corresponding serum specimens. It was later indicated by Laterre et al.,[3] that the duplication of the CSF IgG precipitate in certain CSF specimens of MS patients is due to an imbalance in the CSF κ/λ light chain ratio.

In 1970, Link and Zettervall[4] found an increase in the CSF κ/λ light chain ratio in six out of 11 MS patients in whom oligiclonal gammopathy was demonstrated in CSF agar-gel electrophoresis. It was subsequently recognized that besides individuals with increased concentration of CSF κ chains, a relatively smaller proportion of MS patients may show increased concentration of CSF λ light chains[5].

In this communication, we will demonstrate and discuss the incidence of doubling in the CSF IgG precipitate observed on CSF immunoelectrophoretic examination in 5,203 consecutive patients with various neuropsychiatric disorders.

MATERIALS AND METHODS

In 5,203 consecutive patients hospitalized at the Indiana University Hospitals, the modified Scheidegger's[6] microimmunoelectrophoresis with simultaneous examinations of concentrated CSF and the corresponding serum specimen were performed as previously described[7]. Antisera to human serum and to Fab fragments of IgG (Fab IgG) (Behring Diagnostics, Somerville, NJ; Meloy Laboratories, Springfield, VA) were used. In addition, antisera to κ and λ light chains were applied in 983 patients in whom MS, inflammatory or lymphoproliferative central nervous system (CNS) afflictions were suspected. Electroimmunodiffusion technique[8] for examinations of CSF IgG and albumin concentrations was applied.

RESULTS

The incidence of doubling in the CSF precipitates of Fab IgG and/or IgG is presented in Table 1. Approximately 80% of patients with the CSF immunoelectrophoretic abnormalities studied were individuals with MS. Inflammatory CNS afflictions were diagnosed in an additional 10% in this series. Table 2 shows the incidence of elongation in κ and λ light chains and of doubling in the precipitates of Fab IgG and/or IgG in individuals in whom CSF IgG concentrations were simultaneously determined.

It is obvious that elongation in the precipitate of CSF κ and λ chains, particularly of κ chains, may be demonstrated on routine CSF examination more frequently than doubling in the CSF precipitate of Fab IgG and/or IgG. Doubling in the precipitation arcs of Fab IgG and/or IgG, in absence of demonstrated elongation in the CSF precipitate of light chains, was associated with highest incidence of in-

Table 1. Doubling of the Precipitation Arc of the Fab Fragments of IgG and/or IgG

	Number of Patients	Percentage
Multiple Sclerosis	92	77.9
Multiple Sclerosis - Suspected	5	4.2
Meningoencephalitis	3	2.5
Guillain-Barre' Syndrome	3	2.5
Subacute Sclerosing Panencephalitis	2	1.6
Cerebrovascular Disease	2	1.6
Hydrocephalus	2	1.6
Neurosyphilis	1	0.8
Intracranial Vasculitis	1	0.8
Lymphoma	1	0.8
Inadequate Medical Documentation	6	5.0
TOTAL	118	99.3

Table 2. Cerebrospinal Fluid

Immunoelectrophoresis	Number of Patients	Patients with Increased IgG Concentration (over 14% of Total Proteins)	Patients with Increased Total Proteins Over 50 mg%
Elongation of the precipitate of:			
Kappa light chains	120	79 (65.8%)	42 (35.0%)
Kappa and Lambda light chains	60	29 (48.3%)	26 (43.3%)
The Kappa light chains and doubling in the precipitation arc of the Fab fragments of IgG and/or IgG	28	19 (67.8%)	11 (39.2%)
The Kappa and Lambda light chains and doubling in the precipitation arc of the Fab fragments of IgG and/or IgG	8	6 (75.0%)	4 (50.0%)
Doubling of the precipitate of the Fab Fragments of IgG and/or IgG	13	12 (92.3%)	4 (30.7%)

creased concentration in CSF IgG (92.3%) and with lowest occurrence of elevated CSF total proteins (30.7%). In over 20% of the patients in our series, doubling in the CSF precipitate was demonstrated only by using the antiserum to Fab fragments of IgG. Doubling in the CSF IgG precipitate may be demonstrated in some MS patients (Figure 1, A #1) in the absence of corresponding abnormalities in the precipitate of CSF Fab IgG. The CSF IgG precipitate may show elongation reflecting imbalance in the concentration of CSF light chains (Figure 1, A #2). Doubling in the CSF precipitates of Fab IgG and/or IgG may also be found in the absence of increased concentration of CSF gamma glycoproteins (Figure 1, B #3) which are frequently seen in patients with MS. In MS patients, the most frequent immunoelectrophoretic CSF abnormality found[7] is elongation of the precipitate of κ light chains (Figure 1, C #4). Less frequently, elongation of the CSF precipitates of λ light chains (Figure 1, E #5) compared to the simultaneously obtained precipitation arc of κ light chains (Figure 1, D) may be found. In some MS patients, the precipitate of light chains may form a second arc in the anodal segment (Figure 1, F #6).

DISCUSSION

In our CSF diagnostic laboratories, we have performed CSF immunoelectrophoresis using antisera to human serum and to Fab fragments of IgG in over 15,000 patients with various neuropsychiatric disorders. As was shown on reviewing CSF findings in 5,203 consecutive patients in our series, doubling in the CSF precipitates of CSF Fab and/or IgG was most frequently established in individuals with MS. We agree with the conclusion of Laterre et al.,[3] that the second precipitation arc, parallel to IgG, reflects increased concentration and/or imbalance in CSF light chains.

The precipitate of κ or λ chains may be seen in the CSF immunoelectropherogram as a completely separated, parallel arc. It may also cross the IgG or essentially the γ heavy chain precipitation arc producing a splitting and/or elongation in the anodal or cathodal segment of the IgG precipitate. The λ chain precipitate shows a tendency to be more noticeably elongated in the anodal direction as compared to the precipitation arc of the κ chain. In our patients with MS, we were unable to demonstrate the deconfiguration in IgG and/or γ heavy chain precipitate that is seen in serum and CSF immunoelectrophoresis in patients with malignant lymphomas. It indicates presence of paraproteins which reflect qualitative changes in the γ heavy chain.

Optimal results of CSF immunoelectrophoresis depend greatly on the amount of CSF protein obtained following concentration of the CSF specimen. In patients with doubling in the CSF precipitates of Fab IgG and/or IgG, the immunoelectrophoretic abnormality may be demonstrated even in instances where there is suboptimal quantity of the CSF protein used for immunoelectrophoretic examination.

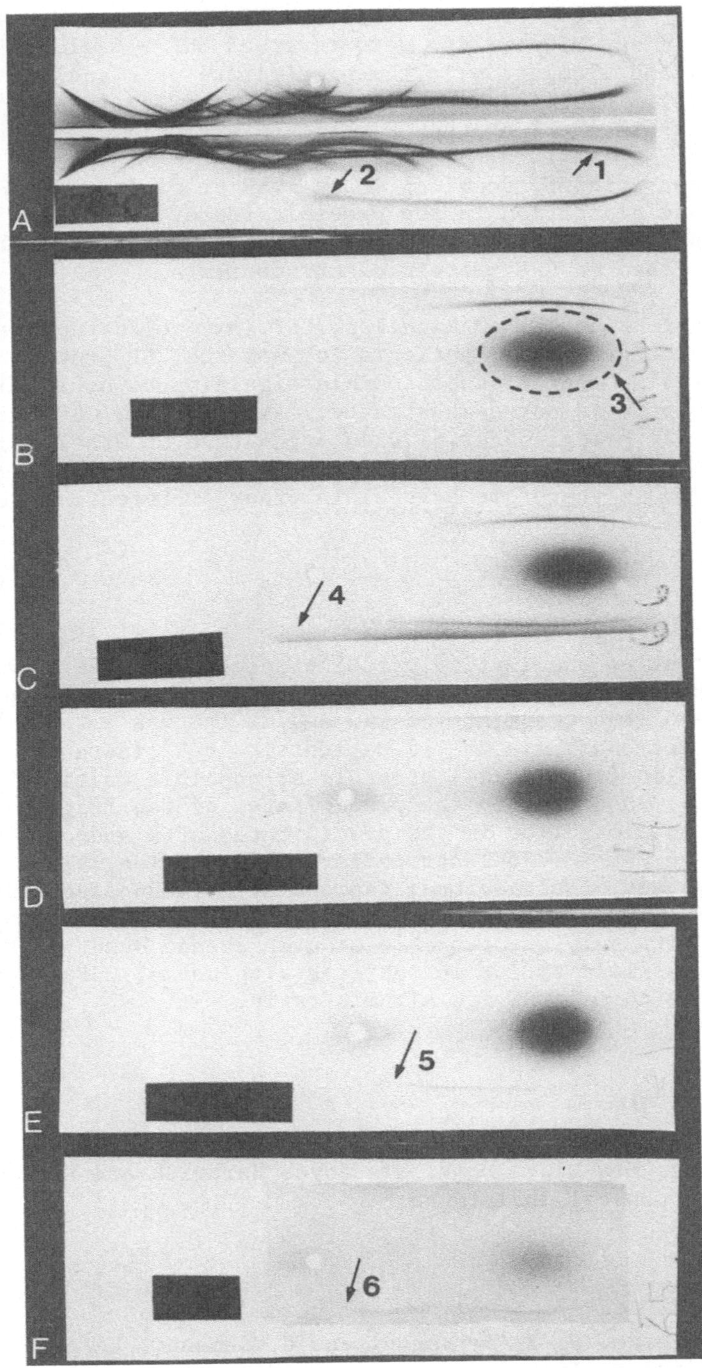

Fig. 1. Cerebrospinal Fluid Immunoelectrophoretic Findings in Multiple Sclerosis.

As we reported previously[7] the most frequent CSF abnormalities found in our MS patients were reversed CSF γ/β globulin ratio (≥ 1; 96.8%) and elongation in the precipitate of κ and/or λ light chains (95.7%). Doubling in the precipitate of Fab IgG and/or IgG was found only in 42.1% of the MS patients examined.

In our opinion, doubling in the CSF precipitate of Fab IgG and/or IgG and elongation in the precipitates of κ and/or λ light chains is an immunoelectrophoretic equivalent of oligoclonal gammopathy established on CSF protein electrophoresis.

As a routine CSF examination, CSF immunoelectrophoresis is particularly helpful in MS patients in whom the CSF protein electrophoresis reveals bands of questionable significance or in whom oligoclonal gammopathy is not demonstrated. In addition, CSF immunoelectrophoresis may reveal abnormalities suggestive of other pathologic mechanisms, as for example, alterations in the blood/CSF barrier structures and is therefore helpful in other differential diagnostic considerations.

SUMMARY

On routine examination of CSF specimens in 5,203 consecutive patients with various neuropsychiatric afflictions, doubling in the precipitates of Fab fragments of IgG and/or IgG was established in CSF immunoelectrophoresis of 118 patients. In 92 instances (77.9%), definite, and in five (4.2%), probable or possible multiple sclerosis was diagnosed. Doubling in the precipitates of Fab fragments of IgG and/or IgG and elongation in the precipitates of κ and/or λ light chains seen on CSF immunoelectrophoretic examination is considered to be the equivalent of oligoclonal gammopathy. Immunoelectrophoretic abnormalities on CSF examination in MS patients may be established in individuals with only questionable or even absent bands in the CSF gamma globulin field and/or in subjects with normal CSF IgG concentration and/or normal CSF IgG/Albumin ratio.

Acknowledgements

The authors thank Mrs Shirley Finchum, Mrs Marcia McClain, Mrs Diana Albright, Mrs Linda Monk, Mrs Joyce Hardwick and Miss Leann Allison for technical assistance.

REFERENCES

1. E. C. Laterre, J. F. Heremans, and G. Demanet, La pathologie protéins du liquide céphalo-rachidien. Etude électrophorétique et immuno-électrophoretique (600 observations). Rev. Neurol., 107:500 (1962).

2. S. J. Dencker, Immuno-electrophoretic investigation of cerebrospinal fluid γ-globulins in multiple sclerosis, Act.Neurol. Scand., 40 (Suppl 10):57 (1964).
3. E. C. Laterre, A. Callewaert, J. F. Heremans, and Z. Sfaello, Electrophoretic morphology of gamma globulins in cerebrospinal fluid of multiple sclerosis and other diseases of the nervous system, Neurol.(Minneap.) 20:982 (1970).
4. H. Link and O. Zettervall, Multiple sclerosis: disturbed kappa: lambda chain ratio of immunoglobulin G in cerebrospinal fluid, Clin.exp.Immunol., 6:435 (1970).
5. O. Kolar and E. Anthony, Cerebrospinal fluid and serum light polypeptide chains in 160 patients with various nervous system disorders, Z.Neurol., 200:6 (1971).
6. J. J. Scheidegger, Une micro-méthode de l'immunoélectrophorèse, Int.Arch.Allergy, 7:103 (1955).
7. O. J. Kolar, P. H. Rice, F. H. Jones, R. J. Defalque, and J. Kincaid, Cerebrospinal fluid immunoelectrophoresis in multiple sclerosis, J.Neurol.Sci., 47:221 (1980).
8. S. A. Schneck and H. H. Claman, CSF immunoglobulins in multiple sclerosis and other neurologic diseases, Arch.Neurol.(Chic.) 20:132 (1969).

MULTIPLE SCLEROSIS: AN ABIOTROPHY WITH HEURISTIC IMPLICATIONS

E. J. Field

Crossley House Neurological Research Centre
17 Brighton Grove
Newcastle upon Tyne NE4 5NS

"The normal process of acceptance of a scientific idea is in four stages.
(i) this is worthless nonsense
(ii) this is an interesting, but perverse point of view
(iii) this is true but quite unimportant
(iv) I have always said so."

J.B.S. Haldane, 1963, J.Genetics, 58:464

"Unless the lay public are taught to appreciate the potential significance of such symptoms as the temporary weakness in a limb, transient dimness of vision, and diplopia the latter part of Buzzard's (1897) criticism 'the full-grown disease is frequently not recognized, the infant disease practically never' must remain justified."

D.K. Adams et al., 1950, Brit.Med.J., ii:431

INTRODUCTION

These words remain as cogent today as when penned; neither is it widely appreciated that 0.5-1.0% of all MS patients are first referred to a psychiatrist (personal communication from Dr. L.H. Field). The existence of totally "silent" cases of MS is now well established (Georgi, 1961; Vost et al., 1864; Ghatek et al., 1974; Morariu and Klutzow, 1976; Castaigne et al., 1981). Georgi's series is biggest and best known. He found amongst 15,644 autopsies carried out at Basle Institute of Pathological Anatomy, 68 cases of anatomically demonstrable MS, in 12 of which (18%) the disease had not been at all suspected during life. Mackay and Hirano (1967) were of the considered opinion that "perhaps clinically silent multiple sclerosis occurs in about one case for each four definitely diagnosed clinically".

Currently, 20% of cases are thus clinically undiagnosable, 2% being benign and 18% clinically silent and are lost to epidemiological and genetic studies - a problem which Myrianthopoulos (1970) discusses in depth. Every neuropathologist of experience has examples of totally unexpected lesions indistinguishable from classical MS in his collection.

Subjects with "subclinical" MS, i.e. individuals in whom minor trivial discomfort such as transient pins and needles in an arm, heaviness in a leg - ?varicose veins, transient pain behind the eye with minor difficulty in reading print etc., who have never thought themselves to be in any way "ill", are common enough if sought and verified by laboratory testing. But meralgia paraesthetica, somaesthetic migraine, unexplained neuritis, minor carpal and other "tunnel" syndromes may all closely mimic early MS as may, of course, "hysteria". To anticipate: there are many more MS subjects brought out by recently developed laboratory testing than present themselves clinically, just as there are many more sufferers from arthritic degenerative change of the knee joint than ever complain to their doctor. It is here proposed that MS (like osteoarthritis) falls into the category of "Abiotrophic Disease" according to the concept associated with the name of Sir William Gowers (1902). These diseases result from a poorly constituted tissue with insufficient "vitality" to last a full three score and ten years. Consequently, it breaks down, apparently "spontaneously", earlier than do other better constituted tissues. Such conditions would be expected to be primarily genetically determined, although secondary mechanisms may come into play in certain individuals once primary degeneration takes plane, e.g. an EAE mechanism in the case of MS. Aggravating or precipitating factors for "spontaneous" myelin breakdown may be added, e.g. puerperium, stress, influenza, etc. A genetic aetiology brings with it the possibility of establishing the biochemical basis of "lessworthy" nervous tissue (Curtius, 1933) and its rectification at a time when myelin is under active construction (between mid-fetal life and 5 years with great intensity and, thereafter, more slowly up to about 16 years) (Yakovlek and Lecours, 1967).

To enunciate clearly the thesis here presented:- Multiple Sclerosis belongs to the category of degenerative disease subsumed under the name of "abiotrophy". Poorly constituted myelin breaks down in scattered regions of the nervous system well before the time of its customary life span. All grades of poorly constituted myelin exist; minimal breakdowns will go unreported by ordinary people (and even by neuropathologists who look only for outstanding plaques). This new concept develops from a consideration of the phenomenology of MS and new laboratory tests. It brings new hope in the total management of the disease.

Pathogenesis of Multiple Sclerosis

In 1863 Rindfleisch observed that "in the most recently degenerated part of the white matter of the brain careful observation with the naked eye showed a red dot or line (Strich), which corresponded to a cross or obliquely cut engorged blood vessel". The great majority of studies have been made on the distribution of advanced lesions by careful reconstruction of serial sections. Plaques then appear as "strings of pearls" strung out along small venules as they converge to a trunk; or as "fingers" stretching along small veins. Rindfleisch also noted that "the earliest basis (Grund) of the disease is to be seen in changes in individual vessels and their branches" - a close forerunner of the "perlschnurig" description to follow decades later. So clear and striking are Weigert-Pal stained plaques with a central blood vessel, especially for teaching purposes, that more and more beautiful photo-micrographs have been reproduced in volume after volume devoted to the pathology of MS, establishing the disease as primarily perivascular. This view has been reinforced by intensive study of EAE (an undoubted perivascular disease) deriving especial impetus from well conceived and authoritative accounts by Waksman (1959) and others, and the wise caution enjoined by Lumsden (1970) disregarded. It must be said, at once, that pathogenetically MS and EAE are quire different. Of basic significance is the absence of lymphocytes from the earliest stage of demyelination in MS. These cells play no part here in the initiation of myelin breakdown. Hassin (see below) and other careful observers have not recorded participation of lymphocytes in earliest myelin disintegration.

Mature plaques are "centered" on small veins. The earliest lesions are quite different. There is in situ degeneration of myelin; it simply disintegrates. No doubt attaches to this. Personal observation of the earliest changes in plaque formation conforms with the vivid and accurate account given by Hassin (1922, 1937). Hassin depicts (1922) (Figure 2) a longitudinal section of an apparently healthy area stained by Bielchowsky's method and counterstained by the method of Alzheimer-Mann. "The axons show no particular changes but the myelin sheath is tumified, forming a broad swollen band around the well preserved axons. In many places the myelin appears fragmented, broken up into globules and droplets. Under the oil immersion there can be observed, in the myelin, more or less wide meshes giving the myelin a reticulated or fenestrated appearance, as if it consists of numerous holes through which the axons can be distinctly seen". The distended meshwork of the myelin may break down giving local cavities - the so-called "Lücken". If these early regions of changes are Marchi stained and counterstained with Alzheimer-Mann, there are, in addition to the above changes, numerous small black structureless droplets (Elzholz bodies) scattered along the axon and intermingled with the fragments of broken myelin. The myelin fragments themselves may be laminated and may stain brown (Marchi's globules). Hassin (loc. cit.) goes on to state firmly that "swollen, fenestrated myelin, its fragmentation with abundant formation of Elzholz bodies and Marchi globules in the presence of a well preserved axon were found in all the thirteen cases studied outside the patch proper in areas which, with ordinary stains, appeared normal". There was no change in blood vessels within these regions of early degeneration.

Hassin's description of the glial reaction is likewise clear and accurate. Like others before him, he found "the myelin changes, however mild, were always accompanied by marked proliferative changes in the glial tissue which showed either increased fibers filling up spaces between the parallel running swollen nerve fibers, or large glial nuclei, rich in chromatin, invested by a visible, well developed membrane and abundantly supplied with cutoplasm". Hassin echoes Anton and Wohlwill's (1912) observation that no significant (nennenswerte) demyelination takes place without simultaneous, or even prior, activation of the astroglia.

Detailed and accurate descriptions by German neuropathologists bear out Hassin's description. Thus, Peters (1958) writes (p. 537), "Swelling and enlargement (Quellung und Schwellung) of myelin sheaths is for the most part (meist) the earliest visible sign of degeneration". Because of this swelling "the sheath develops a pale mesh-like appearance". He quotes serial section work carried out by Falkiewicz at the behest of Marburg, which showed clearly that the shape and extent of a lesion was not dependent upon blood supply and points out that clear vascular changes (ausgesprochene Gefässveränderungen) are notably absent in very fresh lesions. Figure 20 of Peters' account corresponds precisely with the author's own preparations of very early lesions and with the description given by Hassin.

As not uncommonly in the literature, some of the fundamental and penetrating observations on the early pathogenesis of MS lesions go back to Charcot (1872-73, pp. 187-190). He begins by pointing out that not all the details of the origin of lesions are easily made out in tissue fixed in chromic acid. Indeed, fresh material is advantageous in some respects in that it allows of observations which would certainly pass unnoticed in hardened material. Thus, the former reveals the virtually constant existence of "globules et granulations, d'apparence graisseuse ou médullaire" in considerable numbers and these rapidly disappear when the preparation is allowed to stand for some time in chromic acid. Such fatty globules can sometimes be seen still attached to an axis cylinder already partially denuded of myelin. Charcot goes on (p. 190) to draw the distinction, which need not detain us here, between agglomerations of fatty globules (corps granuleux) without nuclei visible (Fettkornchen Agglomerate) and others with nuclei and cell membrane (Fettkornchen Zellen).

The very well known and commonly reproduced illustration, Figure 10 (p. 191) of Charcot's (Figure 3 in DeJong, 1970, where is is, incidently, not altogether accurately labelled) illustrates the Virchow-Robin's space of a blood vessel distended by "goutelettes graisseuses volumineuses" and also "goutelettes graisseuses, groupés en petits amas disseminées, ça et là dans la preparation, en dehors des vaisseaux". Here we come close to Hassin's description. It is of interest to note that more than 100 years later Mackenzie and Wilson (1966) were also able to demonstrate free fatty globules in the brains of scrapie mice when fresh unfixed material, immediately after death, was embedded in 20% coloured gelatin broth and sections cut at once, on a cryostat. Such fat globules are not to be found in fixed material. These observations were readily confirmed by the author both in mouse and natural scrapie of sheep.

No better summary of the true state of affairs relating to the pathogenesis of MS can be given than the mature summary made by Hassin (1937). "Patches or plaques are not the essential feature of multiple sclerosis but merely one of its phases, a terminal condition."

"The essential feature of multiple sclerosis is the diffuseness of the degenerative process, which in the early stages occurs as foci of demyelination."

"Blood vessels play no part of the genesis of the changes in multiple sclerosis, in which the nerve fibers are attacked primarily and directly, without the intermediation of blood vessels."

Recently an attempt has been made to involve small blood vessels, notably by those who believe that vessel blockage by fat emboli, leading to minute hemorrhages, is at the root of MS lesions. Upon this supposition is based the use of Hyperbaric Oxygen Therapy. There is no doubt that very occasionally minute hemorrhages do occur and may even be accompanied by iron deposits indicating their antemortem formation. But they are rare, and Peters writes, "The regular occurrence and importance of such hemorrhages is not established by the majority of workers who have engaged on the study of MS pathology". He quotes Steiner as saying "Occasional (gelegentliche) hemorrhages should be regarded as without significance (Bedeutungslose) and are apparently associated with the terminal state of the patient". Dawson (1916), whose personal observations are a landmark in British neuropathology, has been quoted by James (1982; 1983) as assigning aetiopathogenetic significance to capillary hemorrhages. Acknowledging the frequence with which Siemerling and Raecke (1911; 1914) reported hemorrhages, Dawson sought them with especial care. He reported "in close relation to the engorgement of the blood vessels, both within <u>and without</u> small hemorrhages have been found. These show, however, no changes, and were looked upon as probably the result of respiratory difficulties before death. The vessel walls showed no changes which would explain the hemorrhages, nor were there any signs of inflammation around them" (p. 650); and continues (loc. cit.) "it is difficult to account for the origin of the sclerotic areas in such primary changes. The absence of an histological evidence of changes in the vessel walls associated with thrombosis or hemorrhages is quite incompatible with a primary vascular lesion in this sense". The same agonal explanation was noted above by Steiner.

Glial cells, both astro and oligodendroglia, may act as macrophages and become converted into gitter cells (Ferraro and Davidoff, 1928; Field, 1957), as Hassin also observed.

It is impossible to leave the question of astroglial activation in early MS changes without paying passing tribute to the outstanding, but long forgotten, contributions made by Rossolimo (1897), Strä huber (1903) and others to glial studies in MS. They also recognized the "pre-malignant" state of astroglia near acute lesions as brought forward once more, many years later by Field et al (1962); Field (1967), which explains the occasional association of MS with glioma where there seems to be a gradual transition of enlarged astrocytes into malignant cells (e.g. Scherer, 1938).

Once local myelophages are formed from glial cells they very soon migrate to the nearest (never very distant) small vein so that one has to search diligently to find a region of early primary degeneration without small blood vessels at its margin already surrounded by even a few macrophages stuffed with Sudan IV staining fat. In addition a few lymphocytes and an occasional plasma cell may be present. As myelophages encompass the venule, the number of lymphocytes increases to produce a "cuffed vessel" with numerous myelophages, which gradually becomes incorporated within the degenerated areas. Eccentric, at first, the blood vessel becomes more central as it is engulfed within the abiotrophic region. By the time an established plaque is seen there is a "central" venule and this is the teaching handed down.

The recent revival of interest in the "normal" tissue described above leads to a reappraisal of the whole nature of MS. It is not primarily a perivascular disease at all. The extensive, stimulating and highly influential work by Waksman (1959), which sees experimental allergic disease as motivated primarily by lymphocytes "targeted" on the organ concerned, is misleading when applied to the beginnings of human MS. For MS is not primarily analogous with EAE (despite the statement to this effect which so often begins scientific articles and applications for Research Grants). Once again it should be emphasized that the earliest changes in MS are found in the total absence of lymphocytes.

After this description of the genesis of a lesion in MS the contrast with EAE may be briefly made. In the latter, lesions are initially perivascular, characterized by increased permeability (Lampert and Carpenter, 1965; Field and Raine, 1966), followed by appearance of mononuclear cells. If these break out, together with exuded plasma, into the parenchyma, then demyelinative changes may occur there, though a picture of actual attack of a mononuclear cell on myelin is very rarely seen (Field and Raine, 1966). On the other hand the myelotoxic factor in plasma is found also in ALS (66% of cases) (Bornstein and Appel, 1965; Hughes and Field, 1967).

MS is on the other hand a generalized disease of the central nervous system. Little attention has been paid to the "apparently normal" white matter between discrete lesions. Field (1967) drew attention to widespread changes far beyond the attention-rivetting plaque, especially activation and enlargement of astroglia cells, but in doing so he was very far from the first. Already for Charcot (1872-3) enlargement of the interstitial elements was "incontestablement" the earliest change in MS and he was followed by Müller (1904), Anton and Wohlwill (1912), Dawson (1916) and Jacob (1969). Doubts were cast upon the genuineness of glial changes because of the possibility of minute lesions in the apparent white matter when chemical analysis reported differences in this seemingly normal tissue. Most recently, however, it has come to be realized that MS is a global disease of the CNS with focal intensification apparently around small veins and more modern methods have been brought to bear on the subject. Arstila et al. (1973) reported abnormally then myelin and an increase of astrocyte lysosomes together with increased lysosomal enzymatic activity in biopsies from cerebral white matter outside plaque regions. In well controlled work, Allen and her colleagues (1981, 1983) have extended this work and have come to endorse the writer's long expressed opinion that "categorization of MS material into plaque, peri-plaque and normal white matter may be arbitrary and may limit our thinking on the pathogenesis of the disease". It certainly has; and, indeed, continues to do so and hinders possible progress in disease prophylaxis (see below).

The Multiple Sclerosis Diathesis

Although the "symptomatology and course of MS as described by Charcot (1886) is so complete that subsequent literature may be considered largely as differentiating only regarding type, frequency and quantitation" (Fog, 1977), Charcot did make one fundamental omission in failing to note familial clustering of the disease, perhaps on account of the relatively small number of cases he saw altogether. He was indeed aware of "la famille neuropathologique", and did not fail to ask about other diseases of the nervous system in the patient's family, recording the amusing incident in his "Leçons du Mardi à la Salpêtrière" in which in reply to his question "Je voudrais savoir de vous enfin, s'il n'y a pas eu dans votre famille quelques personnes malades du système nerveux?, he drew the response "Il y a eu des poètes" (Charcot, 1892, p. 392). In the same volume (p. 179) Charcot declared "le clinicien n'a entre ses mains qu'un épisode, s'il veut se borner à l'étude du malade lei-même et n'embrasse pas l'histoire de la famille entière!"

LABORATORY DIAGNOSIS OF MULTIPLE SCLEROSIS

The E-UFA (Field et al., 1977) and PGE_2 (Field and Joyce, 1977) tests are least demanding. But obsessional attention to detail and scrupulous preparation of glassware without the use of any detergent are prerequisites of success. Plastic material of any sort (e.g. as a saline bottle or tubes or disposable pipettes) must be excluded. One "worker" attempting the E-UFA test told me he could not be bothered with cleaning glassware and was going to use plastic. Indeed he added detergent to his reactants! His negative results were published within a few months in a reputable English journal. Another decided to incubate RBA with LA for 30 minutes at 38°C and gave me (the not very surprising) news that the cells underwent lysis! Seaman et al. (1979a) in their confirmation of the E-UFA test gave reasons why others have failed. A very full account of the practical aspects of E-UFA testing has been recorded (Field, 1983).

Theoretical Basis

Since the fundamental work of Burr and Burr (1929; 1930), Swank (1950; 1970) and Sinclair (1956) (see review by Holman, 1978), it has been recognized that certain polyunsaturated fatty acids (PUFA) are essential for proper development and function. The acids are of two groups, the ω-6 and ω-3, and simple explanations are given in the FAO

Table 1. Laboratory Tests for MS.

Test	Reference
1. MEM-LAD	Field et al. 1974
confirmed double blind by	Jenssen et al. 1976
2. E-UFA	Field et al. 1977
confirmed double blind by	Bisaccia et al. 1977
" " " "	Seaman et al. 1979a
" " " "	Tamblyn et al. 1980
" " " "	Jones et al. 1981
3. PGE_2	Field and Joyce 1977
4. PL-EUFA* (Plasma-EUFA)	Field and Joyce 1982a

* Developed from Seaman et al.'s (1979b) demonstration that electrophoretic properties of RBC were conditioned by ambient plasma. Test admits of diagnosis within 30 minutes.

publication of the United Nations on "Dietary Fats and Oils in Human Nutrition" (Rome, 1977). Suffice it here to say that there are two ω-3 acids of considerable importance present in fish oils especially herring, mackerel and cod liver. They are:-

(i) C22: 6 ω-3 (22 C atoms unbranched in a chain with 6 double bonds, the first beginning 3 C atoms from the CH_3 end of the molecule). Its "trivial" name is clupadonic acid.

(ii) C20: 5 ω-3 (20 C atoms in line with 5 double bonds, the first being 3 C atoms away from the terminal CH_3 group). Its "trivial" name is timnodonic acid.

The important acids for laboratory testing purposes are linoleic acid (LA) C18: 2 ω-6 (9.12 Octodecadienoic acid) and arachidonic acid (AA) C20: 4 ω-6 (5.8.11.14 Icosatetraenoic acid - usually mis-spelled "eicosatetraenoic" - Sinclair, 1982).

These acids, as components of phospholipids, are important constituents of cell membranes and if their mishandling were basic to the pathogenesis of MS, as suggested by Thompson (1966; 1973), they would make the membrane of all cells (both surface and internal) different from those in a non-MS subject. Bolton et al. (1968) showed that thrombocytes from an MS subject suspended in plasma showed different electrophoretic mobility from non-MS platelets. They have also been shown to have decreased linolenate content (Gul et al., 1970) as do MS-RBC (Homa et al., 1980).

Following publication by Millar et al. (1973) that supplementation of the diet with 60 ml per day sunflower seed oil (active principle LA) produced (in a double blind trial) a lesser number of exacerbations, of lesser degree and duration than occurred in a control group treated with placebo, further interest in LA was aroused. Although they do not explicitly say so, it appears that Millar et al. were working on the hypothesis of a simple replacement of linoleic acid in the blood in MS where they believed it to be low. In fact these original data have been challenged by Love et al. (1974) who found that reduced LA content of serum lipids was not specific to MS and occurred in all ill patients with acute non-neurologic disease.

Unsaturated fatty acids (UFA) might regulate the immunosuppressive effect of corticosteroids (Turnell et al., 1973), increased linoleate enhancing immunosuppression. Early experiments with the MEM-LAD test (Mertin et al., 1973) showed that whilst oleic acid had no inhibitory effect upon lymphocyte-antigen reaction, both LA and AA and very marked effects. And so, since LA was easier to work with, the MEM-LAD test was born using the well marked difference in suppressive effect of LA on MS lymphocyte-antigen reaction as compared with non-MS lymphocytes. The test was later greatly extended in scope (Field et al., 1974). In vitro, LA, AA and γ-linolenate all have well marked immunosuppressive effect on sensitized lymphocytes (Field and Shenton, 1975).

The claim by Ring et al. (1974) that LA suppresses in vivo immunological reaction (graft rejection in rats) was vitiated by the fact that LA was found to be irritative when injected intraperitoneally; their control group had been untreated (i.e. free from any treatment stress). Mertin (1974) found prolongation of skin graft in mice after subcutaneous injection of LA (though his controls received saline). Brock and Field (1975), however, found the subcutaneous injection of LA was highly toxic in mice; indeed, controls given oleic acid also died. "Graft survival was prolonged in the single mouse which survived LA treatment (16 days as compared with 11.2 ± 0.21 for control group)". Thus toxicity again could be the operative factor through ACTH release. Further experiments with gamma-linolenate (Naudicelle: Bio Oil Research Limited, England) showed that whilst this material was not toxic there was no prolongation of great survival time as compared with controls. This material did not produce a local granuloma as did LA and oleic acid. Brock and Field concluded that these grafting experiments provided no support for the hypothesis that PUFA are significantly involved in immunoregulatory mechanisms.

This point is being dealt with at some length because it is important in the possible mechanism of LA or γ-linolenate action in MS. Further evidence in the human is to hand.

(a) Direct testing of the degree of reactivity of T lymphocytes to PPD by the MEM test was not reduced in subjects who had been taking γ-linolenate for 2 years (Field and Shenton, unpublished).

(b) LA appeared to have a beneficial effect on suppression of a renal graft for up to six months; thereafter rejection did not differ significantly from the control group (McHugh et al., 1977).

We may conclude that whatever be the case in experimental guinea pigs or rats, PUFA have at most a short term cellular immunosuppressive activity in the human. This is an important conclusion respecting its mode of action, if any, in the management of MS. To anticipate, the action of PUFA is believed to be on the structure of cell membranes themselves, via the plasma or tissue fluid in which the cells are living, and not through immunosuppression.

<u>Results of the E-UFA Test</u> (Field et al., 1977)

It must be emphasized that the E-UFA (and other) test was developed for the <u>immediate diagnosis of the young person who comes to his</u>

medical attendant with a suspicious symptom or sign of MS and whose condition has not yet been distorted by therapies such as ACTH, prednisolone etc., etc. A whole gamut of drugs (some of them commonly regarded as innocuous, e.g. $FeSO_4$) have marked tendency to adsorb to the surface of RBC, alter the charge, and so interfere with the result of the E-UFA test. Apart from those mentioned antibiotics, folic acid, very heavy smoking, furadantin, β-blockers (more likely to be met with in elderly relatives of a proband), hygroton (Geigy), prolonged amytal and no doubt many others will interfere, as will sunflower seed oil (LA) and Naudicelle or Efamol (γ-linolenate). Even the taking of soluble asprin within an hour or two of testing can lead to a false result, especially if the more recent plasma method is used. The most careful enquiry must be made about materials bought at Health Stores. Patients, especially from outside England, may come with whole lists of "drugs" they are taking (Field and Joyce, 1981; Field, 1983).

When uncomplicated by drug therapy the E-UFA test shows:-

0.08 mg/ml LA or AA (Sigma London Chemical Company Limited, Poole, England) reduces the electrophoretic mobility of well washed MS-RBC ($p < .001$). RBC from normal subjects or from OND (other destructive neurological diseases) travel more rapidly in the presence of these acids at this concentration.

If 0.02 mg/ml LA or AA is used then these results are reversed: MS-RBC travel more rapidly in the presence of 0.02 mg/ml acids, whilst non-MS RBC travel slowly ($p < .001$).

During the testing of the family of an MS proband "anomalous" subjects will be encountered. All MS mothers and the majority of first born daughters are of this anomalous type. They have not got MS; on repetition of the test several years later the result has been identical; none has developed any suggestion of MS over 6-7 years observation period, but they may give rise to MS offspring (see below).

Results of the PGE_2 Test (Field and Joyce, 1977; 1979)

These have been set out by Field (1980; 1983) and in greater detail by Field and Joyce (1980). The presence of 31.25 pg/ml PGE_2 (made up in medium 199, Hanks based; Gibco Europe, Scotland) causes MS-RBC to travel more slowly than control ($p < .001$); non-MS cells (including those of anomalous subjects) travel more rapidly. At the time of this discovery RBC-receptors for PGE_2 were not known but recently they have been uncovered (Stengel and Hanoune, 1981).

The PGE_2 test is simple and valuable in that if patients have already taken γ-linolenate for about a year they will not give a normal E-UFA test (Field and Joyce, 1978), but the PGE test remains positive until some 20-21 months later (Field and Joyce, 1983).

Family Studies

Whilst as long ago as 1896 Eichorst had recorded MS in a 38 year old mother and 8 year old son, it was Mackay's series of papers and the excellent review of Moya (1962) leading up to the masterly presentation by Myrianthopoulos (1970) which really established the familial element in the disease (though there is a considerable number of papers in the intervening German literature).

Recent work with HLA haplotypes confirms the genetic element in MS though it does little more than tell us there is a diathesis. It may, however, lead to advances in mapping of human chromosomes.

Some 473 English families have been studied in as complete a manner as possible. No obvious Mendelian principles operate in the familial distribution of the disease. Reactivity to LA and AA when followed out in families shows that all mothers of the probands are "anomalous". All of 243 ("English" up to September 1983) mothers of MS probands exhibit such anomaly. (Four additional mothers were themselves suffering from clinical MS and gave a positive E-UFA test.) Here 0.08 mg/ml LA causes RBC to travel more slowly ($p < .001$) - though less so than in MS - whilst 0.08 mg/ml AA causes RBC to travel more rapidly ($p < .001$). Moreover, when such "anomalous" RBC are tested with 0.02 mg/ml LA and AA there is no simple reversal. With 0.02 mg/ml LA the RBC travel more slowly than control (absolute alcohol along added) ($p < .001$); 0.02 mg/ml LA gives the same effect as 0.08 mg/ml LA; but with 0.02 mg/ml AA there is reversal and the cells travel slowly (as with normals) ($p < .001$) (Field, 1980). More than 40% of near relatives, chiefly female, give "anomalous" results. The female descent of the anomally is mindful of mitochondrial inheritance (Fine, 1977; 1978). Both mitochondrial and cytoplasmic plasmic inheritance have not received consideration in MS.

Details of the families have been set out elsewhere (Joyce and Field, 1980; Field, 1980). The E-UFA positivity is found in about 1 in 50 of first degree relatives (parents, brothers, sisters) either as overt or as a "subclinical" or totally "silent" case. The "subclinical" are those with clear symptoms - occasional pins and needles in arms or legs; occasional heaviness in a leg; occasional vision blurring; great tiredness etc., none of which has taken them to their doctor. On casual enquiry, worded so as not to arouse suspicion or anxiety, about half admit to some of the minor symptoms or signs which they attribute to "sleeping on the wrong side in bed", etc. Truly "silent" cases make up the other half. These are people at risk but who may well go through life without any real trouble.

A full analysis of risk amongst relatives is currently under way on clinical grounds, clinical + E-UFA test and relatives "said to have had" MS (based on corroboration).

To take some examples from "English" families only:-

Altogether 275 mothers (of English families) were examined of which 243 were "anomalous" (see above), 4 were clinical open MS, and 17 were "silent" or "subclinical" MS giving minor or suspicious symptoms or signs on oblique questioning; 11 more were "told" to us on good grounds making a total of 4 + 17 + 11 = 32.

Therefore, 32/275 = 11.64% of mothers of an MS proband were themselves MS. If we assume 60 per 100,000 as the prevalence rate in England, then:-

$\frac{100,000}{276}$ x 32 x 1/60 = 194 x as common as in the general population.

There were in addition 4 cases where a silent MS sibling had a silent MS mother.

Turning to fathers of an MS proband: altogether 114 fathers were available for study. Of these 2 were clinical MS and 10 were subclinical.

Therefore, 2/114 = 10.53% of fathers of MS probands were themselves MS. Taking into account those brought out by the E-UFA test we had 12 MS fathers. Again assuming an English prevalence rate of 60 per 100,000 we have $\frac{100,000}{114}$ x 12 x 1/60 = 175.4 x as common as in the general population.

Adding parents together, we get:-

32 + 12 = 44 MS out of 275 + 114 = 389 and 44/389 = 11.31% of parents of an MS proband are themselves MS.

Therefore, $\frac{100,000}{389}$ x 44 x 1/60 = 188.5 x as common amongst parents of probands as in the general population (taken uncorrected) at 60 per 100,000.

Turning now to English siblings we have 99 brothers and 173 sisters available for study who were normal and 47 probands.

	Brothers	Sisters
Clinically MS	2	2
"Silent" (E-UFA)	18	21
"Told"	2	2
	22	25

i.e. 47 MS out of 272 + 47 = 319

therefore 47/319 = 14.73% of siblings of an MS proband are themselves MS, so that:-

$\frac{100,000}{319}$ x 47 x 1/60 = <u>245.6 x as common as in the general population</u>
(reckoned - uncorrected - as 60 per 100,000)

and $\frac{\text{SIBLINGS}}{\text{PARENTS}} = \frac{14.73}{11.31} = 1.3$

i.e. there are 1.3 times as many MS to be found amongst siblings of a proband as amongst the parents.

These figures illustrate how greatly the occurrence of MS is magnified when the E-UFA test is employed.

<u>Percentage of families with more than one case in the family</u>, i.e. a proband + at least one more case; we have on clinical ground alone 8/473 = 1.69%.

If we take those who are E-UFA positive then we have:-

Clinical + "silents" = 104/46 = 21.98%, i.e. percentage goes up by a factor of 13 if we use E-UFA.

If we include another 24 of which we were "told", then:- 128/473 = <u>27.06%</u>.

Other workers who have not used E-UFA testing of all available family members will have used clinicals + "tolds", i.e. 8 + 24 = 32, i.e. 32/473 = <u>6.76%</u> of MS families they will consider to have at least two in the family.

If Falconer's (1965) Figure 4 is used as a rough guide to heritability then our figures suggest an 80-100% heritability in MS. However limited the usefulness of this estimate (which is much above that calculated by Berry (1969) from the figures of other workers) it does allow us to say with confidence that a high genetic element operates in the development of MS.

Several points emerge:-

(a) All mothers of MS patients are "red circles" (in so far as can be proved by "induction"), i.e. their RBC travel more slowly with 0.08 mg/ml LA and fast with 0.08 mg/ml AA.

(b) A high proportion of first daughters of an MS parent are also "red circles" and these, whilst never getting MS themselves, can actually produce MS offspring.

(c) A normal female child can neither herself develop MS, nor become the mother of an MS offspring even if she has the misfortune to marry an MS husband. Family testing does much to relieve scarcely concealed anxieties within MS families.

Much of this work, with some elementary genetic counselling, has been set out elsewhere (Field and Joyce, 1980). Our limited experience with Italian families strongly suggests that Italy and Sicily are high risk areas comparable with England; and about 40% of near relatives are anomalous as in England (see below).

Recurrent Acute Disseminated Encephalomyelitis

This is relatively rare and great attention should be paid to initial clinical history - drowsiness, fever, antecedent respiratory or other infections and so on. Such cases, when they are of the (rare) recurrent type, can readily pass under the heading of MS. However, they give a negative E-UFA test and their mothers are not "red circles".

Neurosarcoidosis, likewise, gives a negative result and the mother, once again, is not a "red circle". It must be remembered that cases of Boeck's sarcoid may be confined to the CNS and give general signs (reviewed by Matthews, 1979). The way in which neurosarcoidosis may imitate MS clinically is set out in great detail by Zollinger (1941).

DISCUSSION

How widespread is MS?

The E-UFA test not only enables diagnosis of MS to be made at the first presenting symptom but uncovers an "MS diathesis" which in the great majority of cases will eventuate in clinical manifestations, though these may be very mild. This diathesis is much more frequent than clinical signs recognized. The same phenomenon appears to occur in all areas looked at, including Italy and Sicily. There are far more cases of MS brought to light than are supposed to occur in this "low risk area". In England the number of cases generally set at about 50,000 is probably nearer 125,000. The Vale of Trent has been especially explored and is rife with MS. Recently, indeed, there have been epidemiological studies in Italy which put up the prevalence and incidence rates on clinical grounds along. Thus, Dean et al. (1981) found "at least 32 per 100,000 in Agrigento city on the south west coast of Sicily" but thought "this is likely to be a considerable underestimate of the true prevalence". Agrigento is given as an example of the difficulty of studying MS in a rural city with no specific interest in neurological problems and far from a center. We would agree that 32/100,000 is an underestimate and E-UFA testing brought to bear upon the problem would assuredly at least double the prevalence when relatives of probands (themselves to be verified by testing!) are studied. We disagree, however, that absence of good neurological facilities make the recognition of MS so much more difficult. Epidemiologists must rid themselves of "possible" cases; there is MS, non-MS and an MS-like syndrome. The application of the E-UFA test to eliminate "possible" MS has recently been outlined (Field and Joyce, 1982b). The major practical difficulty is transport of blood specimens or the setting up of a local testing station. If MS is of the nature of an abiotrophy then we would expect many cases of MS to occur with small and unnoticed lesions during life where disintegration had not gone beyond a few plaques possibly located in non-critical areas.

All these "silent" and subclinical cases will be brought to light by application of the E-UFA test under the proper conditions (outlined above) and make a very considerable difference to the final figure especially in the parent-proband and proband-sibling ratios, and hence on the general frequency of the condition in the population at large.

On the Nature of MS

If the basis of our tests be correct and there is indeed an inborn mishandling of unsaturated fatty acids reflected in the unusual phospholipid component of cell membrane, then it is highly probable that this charge extends to that most membranous organ of the body - the brain. If the surface of the oligodendrocyte be affected in the same way as the RBC, then myelin as a product in large measure of the oligodendrocyte surface might also be expected to be unusual (Field and Joyce, 1982b). Here we enter into the realms of (testable) hypothesis. It may be supposed that the MS modified oligodendrocyte membrane has different surface groupings which do not exert the full and powerful electrostatic binding between adjacent lamellae. Consequently there is a tendency for myelin lamellae to spring apart - like watch-spring leaves. It may be further supported that certain HLA groupings (known to be present on glial cells, but apparently not on neurones) may also contribute to lamellar-binding: certain others may be weak, or entirely deficient in this respect. Those HLA haplotypes which are of the latter nature may be just those found most frequently in MS. The MS inclined myelin would then be doubly at risk in that the lamellar-binding forces would be deficient from two causes. Thus we may suggest that at root the cause of MS (i.e. the localized, isolated regional breakdown or crumbling of myelin in situ which we call a "plaque") is due to defective constitution and binding of myelin lamellae. MS myelin then is, in the great majority of cases, endowed with inadequate "vitality" to hold out for a natural life span, and we come to the notion of abiotrophy, introduced by Sir William Gowers in 1902. This postulates (in modern molecular biological language) a tissue inadequately programmed - programmed, indeed, for early death. Gowers - whose paper should be read by all interested in medicine and not only neurology uses more homely (and equally enlightening) terms - "lack of vitality" in the tissues. MS according to this view comes into the same category as osteoarthritis of the knee joint and is probably just as common. Certainly the range of affliction in the latter varies from "normal wear and tear" at over 60 years through to severe degeneration changes as early as 30 years. Many people with minor creaks and twinges in the knee never bother their doctor. The case is similar with MS.

All degrees of degeneration are to be found with little or no correlation between plaque distribution and symptoms. Many MS sufferers never go to the doctor and those who do may be symptomatically treated without the true nature of the complaint being recognized.

There is, however, one respect in which osteoarthritis and MS differ markedly. It is possible to pick up the tell-tale marker of MS soon after birth by the E-UFA test whilst the future osteoarthritic cannot be so diagnosed. The analogy is closer to phenylketonuria, again a programmed disease but one in which the precise biochemical defect is known, can readily be ascertained, and can be rectified at once.

If the argument regarding MS has been followed, then it will appear that once abnormal cell surface has been diagnosed, measures should be taken to rectify it and return it, at least so far as electrophoretic properties (and indeed biochemical constitution) (Homa et al., 1981) are concerned, towards normal. This can easily be done by administering LA, or better, γ-linolenate, in the young child as in the adult. Clearly it is important that this rectification be carried out as early as possible and certainly whilst myelination is active, i.e. from mid-fetal life until the age of about 16 years, although there is evidence (Jungawala and Dawson, 1971; Horrocks, 1973; Norton and Podulso, 1973; Dawson and Gould, 1976; Gould, 1977) that in rats some myelin constituents undergo metabolism till relatively late in life. Myelin cannot be considered a discrete metabolic entity, but has many components, with different turnover rates so that it may be possible to influence its composition well into adult life.

There appears to be a continual increase in the amount of myelin in the brains of mice and rats throughout the animal's lifetime (Horrocks, 1973). Norton and Podulso (1973) have commented upon the extraordinarily large amount of myelin produced by each oligodendrocyte. For the rat they make the rough calculation that each oligodendrocyte is making 175×10^{-12} gm of myelin each day or about threefold its own weight.

Although the older morphological studies of Yakovlev and Lecours (1967) suggested that active myelination in the human was going on until the age of 5 years and thereafter with diminishing intensity until about 16, other studies by Brante (1949), Tingey (1956) and others suggest that by 2 years of age 90% of the total lipid has been deposited and that by about 10 years adult values have been attained. However, the question of turnover or replacement of individual components remains an unsettled question and leaves open the possibility of replacement therapy.

The EAE Mechanism in MS

Research should be directed to picking out individuals in whom an EAE mechanism has been superadded to the primary breakdown of myelin. If this can be demonstrated, e.g. by ratios of lymphocyte subspecies, then immunosuppression would be a rational treatment. However, it is clearly not initiated by brain trauma. Recurrent ADEM, too, is rationally treated with immunosuppressants.

We would appear to be at a turning point in our whole view of MS and an optimistic road lies ahead. But, as Sir Harold Himsworth has pointed out (1949), "The history of modern knowledge is concerned in no small degree with man's attempt to escape from his previous concepts". The view of MS to which we have been forced is, of course, by no means universally accepted; it is, however, to be hoped that for some thoughtful workers it has, at least, passed from the late Professor Haldane's stage (a) into stage (b).

APPENDIX

Amongst English families the "anomalous" E-UFA result is found in some 42% of near relatives of an MS index case. There is 100% occurrence amongst mothers, very low amongst fathers and other male relatives, but it is 66% amongst sisters, 63.8% amongst daughters and 33% amongst nieces.

Our Italian figures are much smaller, for geographic reasons, as soon from Table 2. This reinforces strongly our impression that MS is every bit as common in Italy (both north and south, including Sicily) as it is in England and makes systematic study of well defined areas more urgent. Clearly a rectification of prevalent ideas on geographic distribution must be important in solving MS, and we would hazard a suggestion that it has more to do with the wanderings of the Northmen and the distribution of their genes than with latitude.

Table 2. "Anomalous" Relatives (i.e. slow with LA, fast with AA)

	Number	"Anomalous"	%
mothers	15	15	100
fathers	7	0	0
sons	17	1	5.8
daughters	12	8	66.7
sisters	12	8	66.7
brothers	8	3	37.5
others	17	5	29.4
	88	40	45.0%

Acknowledgements

The author would like to express his gratitude to Miss Greta Joyce for the major part of the practical work upon which these ideas were based. The work has been supported throughout by the Naomi Bramson Trust.

ADDENDUM

Myalgic encephalitis (ME) may on occasion have to be considered in the differential diagnosis of early MS. Since the above work was completed we have had the opportunity (through the courtesy of Dr. D.G. Smith, Medical Advisor to the ME Association) to study several classical examples of the disease. This has been made with a new (as yet unpublished laser method) which separates MS, OND and normal subjects. Those with ME fall into the OND category and can easily be set aside from MS.

REFERENCES

Allen, I.V., 1983, Hydrolytic Enzymes in MS, *in*: "Progress in Neuropathology", Vol. 5, H.M. Zimmerman, ed., Raven Press, New York, 1-17.

Allen, I.V., Glover, G., and Anderson, R., 1981, Abnormalities in the microscopically normal white matter in cases of mild or spinal Multiple Sclerosis (MS)., *Acta Neuropath.(Biol)*., Suppl. IV:176-178.

Anton, G., and Wohwill, F., 1912, Multiple nicht eitrige Encephalomyelitis und Multiple Sklerose., *Z.ges.Neurol.Psychiat*., 12:31-98.

Arstila, A. U., Riekkinen, P., Rinne, U. K., and Laitinen, L., 1973, Studies on the pathogenesis of multiple sclerosis: Participation of lysosomes on demyelination in the central nervous system while matter outside plaques., Eur.Neurol., 9:1-20.

Berry, R. J., 1969, Genetical factors in the aetiology of Multiple Sclerosis, Acta Neurol.Scand., 45:459-483.

Bisaccia, G., Caputo, D., and Zibetti, A., 1977, E-UFA test in Multiple Sclerosis, Boll.Ist.sieroter Milan., 56:583-588.

Bolton, C. H., Hampton, J. R., and Phillipson, O. T., 1968, Platelet behaviour and plasma phospholipids in MS., Lancet., 1:99-1C4.

Bornstein, M. B., and Appel, S. H., 1965, Tissue culture studies cf demyelination, Ann.NY Acad.Sci., 122:280-286.

Brante, G., 1949, Studies on lipids in the nervous system with special reference to qualitative chemical determination and topical distribution, Acta Physiol.Scand., (Suppl. 63), Stockholm, 18:1-20.

Brock, J., and Field, E. J., 1975, Unsaturated Fatty Acids and Transplantation, Lancet., 1:1382-1383.

Burr, G. O., and Burr, M. M., 1929, A new deficiency disease produced by the rigid exclusion of fat from diet, J.Biol.Chem., 82: 345-367.

Burr, G. O., and Burr, M. M., 1930, On the nature and role of the fatty acids essential in nutrition, J.Biol.Chem., 86:587-621.

Castaigne, P., Lhermitte, F., Escourolle, R., Huan, J. J., Gray, F., and Lyon-Caen, O., 1981, Les sclèroses en plaques asymptomatiques, Rev.Neurol.(Paris), 1937:729-739.

Charcot, J-M., 1872-1873, Lecons sur les maladies du système nerveux. Bournville, ed., Delahaye, Paris, 187-190.

Charcot, J-M., 1886, Lecons sur les maladies du système nerveux, in: "Oeuvres complètes de J-M Charcot", Tome 1, Bournville, ed., Paris, 158-163.

Charcot, J-M., 1892, Lecons du mardi à la Salpêtrière, Vol. 1, Bataille, Paris, 392.

Curtius, F. C., 1933, "Multiple Sklerose und Erbanlage", Georh Thieme Verlag, Leipzig, 215.

Dawson, J. W., 1916, The jistology of disseminated sclerosis, Trans.Roy.Soc.Edin., 50:517-740.

Dawson, R. M. C., and Gould, R. M., 1976, Renewal of phospholipids in the myelin sheath, Adv.Exp.Med.Biol., 72:95-113.

Dean, G., Savettieri, G., Tairi, G., Morreale, S., and Karhausen, L., 1981, The prevalence of multiple sclerosis in Sicily, II Agrigento City, J.Epidemiol.Commun.Health., 35:118-122.

DeJong, R. N., 1970, Multiple Sclerosis, History definition and general consideration, in: "Handbook of Clinical Neurology", P.J. Vinken and G.W. Bruyn, eds., North Holland Publishing Co., Amsterdam and New York, 9:45-63.

Eichorst, H., 1896, Ueber infantile und hereditäre multiple Sklerose, Archiv f.Pathol.Anat., 146:173-192.

Falconer, D. S., 1965, The inheritance of liability to certain diseases, estimated from the incidence among relatives, Ann.Hum.Genet.Lond., 29;51-76.

Ferraro, A., and Davidoff, L. M., 1928, The reaction of the oligodendroglia to injury of the human brain, Arch.Path., 6:1030-1063.

Field, E. J., 1957, Histogenesis of compound granular corpuscles in the mouse brain after trauma and a note on the influence of cortisone, J.Neuropath.Exp.Neurol., 16:48-56.

Field, E. J., 1967, The significance of astroglial hypertrophy in scrapie, kuru, multiple sclerosis and old age, together with a note on the possible nature of the scrapie agent, Dtsch Z. Nervenheilk., 192:265-274.

Field, E. J., 1980, "Multiple Sclerosis in Childhood: Treatment and Prophylaxis", C.C. Thomas, Springfield, Illinois.

Field, E. J., 1983, E-UFA Test, in: "Cell Separation Methods and Selected Application", Vol. 2, T.G. Pretlow II and T.P. Pretlow, eds., Academic Press, New York and London, 251-271.

Field, E. J., and Joyce, G., 1977, Prostaglandin (PGE_2) and human erythrocyte mobility: a specific test for multiple sclerosis? IRCS Med.Sci., 5:158-159.

Field, E. J., and Joyce, G., 1978, Effect of prolonged ingestion of gamma-linolenate by MS patients, Eur.Neurol., 17:67-76.

Field, E. J., and Joyce, G., 1979, Multiple Sclerosis: what can and cannot be done, Brit.Med.J., 2:1571-1572.

Field, E. J., and Joyce, G., 1980, Recent Methods of in vitro Diagnosis of Multiple Sclerosis, in: "Progress in Multiple Sclerosis Research", H.J. Bauer, S. Poser and G. Ritter, eds., Springer-Verlag, Berlin and Heidelberg, 310-322.

Field, E.J., and Joyce, G., 1981, Simplified E-UFA test for Multiple Sclerosis (MS): Some sources of "false" results, J.Neurol., 226:149-155.

Field, E. J., and Joyce, G., 1982a, An Office Test for Multiple Sclerosis, IRCS Med.Sci., 10:560-561.

Field, E. J., and Joyce, G., 1982b, A laboratory test to achieve assured ascertainment of MS; frequency in epidemiological studies and clinical trials, in: "Multiple Sclerosis East and West", Y. Kuroiura and L.T. Kurland, eds., Kyushu University Press, Fukuoka, Japan, 265-272.

Field, E. J., and Joyce, G., 1983, Multiple Sclerosis: Effect of gamma-linolenate administration upon membranes and the need for extended clinical trials of unsaturated fatty acids, Eur.Neurol., 22:78-83.

Field, E. J., and Raine, C. S., 1966, Experimental allergic encephalomyelitis: an electron microscope study, Amer.J.Path., 49:537-553.

Field, E. J., and Shenton, B. K., 1975, Inhibitory effect of unsaturated fatty acids on lymphocyte-antigen interaction with special reference to Multiple Sclerosis, Acta Neurol.Scand., 52:299-309.

Field, E. J., Miller, H., and Russell, D. S., 1962, Observations on glial inclusion bodies in a case of acute disseminated sclerosis, Brit.J.Clin.Path., 15:278-284.

Field, E. J., Shenton, B. K., and Joyce, G., 1974, Specific laboratory testing for diagnosis of Multiple Sclerosis, Brit.Med.J., 1:412-414.

Field, E. J., Joyce, G., and Smith, B. M., 1977, Erythrocyte-UFA (E-UFA) mobility test for Multiple Sclerosis: Implications for pathogenesis and handling of the disease, J.Neurol., 214: 113-127.

Fine, P. E. M., 1977, Analysis of family history data for evidence of non-Mendelian inheritance resulting from vertical transmission, J.Med.Genet., 14:399-407.

Fine, P. E. M., 1978, Mitochondrial inheritance and disease, Lancet, 2:659-662.

Fog, T., 1977, Clues from clinical features, in: "Multiple Sclerosis: A Critical Conspectus", E.J. Field, ed., MTP Press Ltd, Lancaster, England, 13-33.

Georgi, W., 1961, Multiple Sklerose, Pathologischanatomische Befunde multipler Sklerose bei klinisch nicht diagnostizierten Krankheiten, Schweiz.Med.Wochenschr., 91:606-607.

Ghatak, N. R., Hirano, A., Litjmaer, H., and Zimmerman, H. M., 1974, Asymptomatic demyelinated plaque, Arch.Neur.(Chic)., 30:484-486.

Gould, R. M., 1977, Incorporation of glycoproteins into peripheral nerve myelin, J.Cell Biol., 75:306-338.

Gowers, W., 1902, A lecture on abiotrophy, Lancer, 1:1003-1007.

Gul, S., Smith, A. D., Thompson, R. H. S., Payling Wright, H., and Zilkha, K. J., 1970, The fatty acid composition of phospholipids from platelets and erythrocytes in Multiple Sclerosis, J.Neurol.Neurosurg. Psychiat., 33:506-510.

Hassin, G. B., 1922, Studies in the pathogenesis of Multiple Sclerosis, Arch.Neurol.Psychiat., 7:589-607.

Hassin, G. B., 1937, Pathologic features of multiple sclerosis and allied conditions, Arch.Neurol.Psychiat., 38:713-724.

Himsworth, H. P., 1949, The syndrome of diabetes mellitus and its causes, Lancet, 1:465-473.

Holman, R. T., 1978, How essential are essential fatty acids? J.Amer.Oil Chemists Soc., 55:774A-781A.

Homa, S. T., Belin, J., Smith, A. D., Monro, J. A., and Zilkha, K. J., 1980, Levels of linolenate and arachidonate in red blood cells of healthy individuals and patients with multiple sclerosis, J.Neurol.Neurosurg.Psychiat., 43:106-110.

Homa, S. T., Conroy, D. M., Belin, J., Smith, A. D., and Zilkha, K. J., 1981, Fatty acid patterns of red blood cell phospholipids in patients with multiple sclerosis, Lancet, 2:474.

Horrocks, L. A., 1973, Compounds and metabolism of myelin phosphoglycerides during maturation and ageing, Prog.Brain Res., 40:383-387.

Hughes, D., and Field, E. J., 1967, Myelotoxicity of serum and spinal fluid in MS: a critical assessment, Clin.Exp.Immunol., 2:295-309.

Jacob, H., 1969, Tissue process in MS and parainfections and post-vaccinal encephalomyelitis, Int.Arch.Allergy, 36:suppl. 22-36.

James, P. B., 1982, Evidence for subacute fat embolism as the cause of Multiple Sclerosis, Lancet, 1:380-386.

James, P. B., 1983, Oxygen for Multiple Sclerosis, Lancet, 2:632.

Jenssen, H. L., Meyer-Rienecker, H. J., Köhler, H., and Günther, J. K., 1976, The Linoleic Acid Depresseion (LAD) Test for Multiple Sclerosis using the Macrophage Electrophoretic Mobility (MEM) Test, Acta Neurol.Scand., 53:51-60.

Joyce, G., and Field, E. J., 1980, Further observations with the Erythrocyte-Unsaturated Fatty Acid Test. A contribution to the Genetics of Multiple Sclerosis, Eur.Neurol., 19:266-272.

Jones, R., Capildeo, R., Rose, C. F., Forrester, J. A., Luckman, N. P., and Preece, A. W., 1981, A diagnostic test for multiple sclerosis using glutaraldehyde fixed erythrocytes and laser cytopherometry, in: "Cell Electrophoresis in Cancer and Other Clinical Research", A.W. Preece and P. Ann Light, eds., Elsevier/North Holland Biomedical Press, Amsterdam, New York, Oxford, 189-195.

Jungawala, F. B., and Dawson, R. M. C., 1971, The turnover of myelin phospholipids in the adult developing rat brain, Biochem.J., 123:683-693.

Lampert, P., and Carpenter, S., 1965, Electron microscope studies on the vascular permeability and the mechanism of demyelination in experimental allergic encephalomyelitis, J.Neuropath.Exp. Neurol., 24:11-24.

Love, W. C., Cashell, A., Reynolds, M., and Callaghan, N., 1974, Linoleate and fatty acid patterns of serum lipids in multiple sclerosis and other disease, Brit.Med.J., 2:18-21.

Lumsden, C. E., 1970, The neuropathology of MS, in: "Handbook of Clinical Neurology", P.J. Vinken and G.W. Bruyn, eds., North Holland Publishing Co., Amsterdam and New York, 9:217-309.

Mackay, R. P., and Hirano, A., 1967, Forms of benign Multiple Sclerosis, Arch.Neurol.(Chic.), 17:588-600.

Mackenzie, A., and Wilson, A. M., 1966, Accumulation of fat in the brains of mice affected with scrapie, Res.Vet.Sci., 7:45-54.

Matthews, W. B., 1979, Neurosarcoidosis, in: "Handbook of Clinical Neurology", P.J. Vinken, G.W. Bruyn and H. Klawans, eds., North Holland Publishing Co., Amsterdam and New York, 38, part 1:521-542.

McHugh, M. I., Wilkinson, R., Elliott, R. W., Field, E. F., Dewar, P., Hall, P. R., Taylor, R. M. R., and Uldall, P. R., 1977, Immunosuppression with polyunsaturated fatty acids in renal transplantation, Transplantation, 24:263-267.

Mertin, J., 1974, Unsaturated fatty acids and renal transplantation, Lancet, 2:717.

Mertin, J., Field, E. J., and Caspary, E. A., 1973, Unsaturated fatty acids in multiple sclerosis, Brit.Med.J., 2:777.

Millar, J. H. D., Zilkja, K. J., Langman, M. J. S., Payling Wright, H., SMith, A. D., Belin, J., and Thompson, R. H. S., 1973,

Double blind trial of linoleate supplementation of the diet in multiple sclerosis, Brit.Med.J., 1:756-768.

Morariu, M., and Klutzow, W. F., 1976, Subclinical Multiple Sclerosis, J.Neurol., 213:71-76.

Moya, G., 1962, Sclérose en plaques familiales. Etude critique de leur signification, Acta Neurol.et Psychiat.Belg., 62:40-94.

Müller, E., 1904, Die multiple Sklerose des Gehirns und Rückenmarks, Verlag von G. Fisher, Jena.

Myrianthopoulos, N. C., 1970, Genetic aspects of multiple sclerosis, in: "Handbook of Neurology", P.J. Vinken and G.W. Bruyn, eds., North Holland Publishing Co., Amsterdam and New York, 9:85-106.

Norton, W. T., and Podulso, S. E., 1973, Myelination in rat brain. Changes in Myelin composition during brain maturation, J.Neurochem., 21:759-762.

Peters, G., 1958, Multiple Sklerose, in: "Handbuch der Spezellen Pathologischen Anatomie und Histologie", P. Lubarsh, F. Henke, and R. Rössle, eds., Springer-Verlag, Berlin, Göttingen and Heidelberg, 13 part II:525-602.

Rindfleisch, E. E., 1863, Histologisches Detail zu der graven Degeneration von Gehirn und Rückenmark, Virchows Arch., 26:474-483.

Ring, J., Seifert, J., Mertin, J., and Brendel, W., 1974, Prolongation of skin allografts in rats by treatment with linoleic acid, Lancet, 2:1331.

Rossolimo, G., 1897, Zur Frage über die multiple Sklerose und Gliose, Deutsch.Zeitschr.f.Nervenheilk., 11:89-121.

Scherer, H. J., 1938, La "Glioblastomatose en plaques" sur les limites anatomiques de la gliomatose et des processus sclérotique progressifs (sclérose en plaques, sclérose diffuse de Schilder, sclérose concentrique), J.Belge Neurol., 38:1-17.

Seaman, G. V. F., Swank, R. L., Tamblyn, C. H., and Zukoski, C. F.iv, 1979A, Simplified red cell electrophoretic mobility test for multiple sclerosis, Lancet, 1:1138-1139.

Seaman, G. V. F., Swank, R. L., and Zukoski, C. F.iv, 1979b, Red-Cell membrane differences in multiple sclerosis are acquired from plasma, Lancet, 1:1139.

Siemerling, E., and Raecke, J., 1911, Zur pathologischen Anatomie und Pathogenese der multiplen Sklerose, Arch.f.Psychiat.u. Nervenkr., 43:824-840.

Siemerling, E., and Raecke, J., 1914, Beitrag zur Klinik und Pathologie der multiplen Sklerose mit besonderer Berücksichtigung ihrer Pathogenese, Arch.f.Psychiat.u. Nervenkr., 53:385-564.

Sinclair, H. M., 1956, Deficiency of essential fatty acids and atherosclerosis, etcetera, Lancet, 1:381-383.

Sinclair, H. M., 1982, Icosapentaenoic acid and ischaemic heart disease, Lancet, 2:383.

Stengel, D., and Hanoune, J., 1981, The catalytic unit of ram sperm adenylate cyclase can be activated through the guanine nucleotide regulatory component and prostaglandin receptors of human erythrocyte, J.Biol.Chem., 256:5394-5398.

Strähuber, A., 1903, Ueber Degenerations - und proliferationsvorgänge bei multipler Sklerose des Nervensystems, Zieglers Beitr.zu Path.Anat., 33:409-480.

Swank, R. L., 1950, Multiple Sclerosis: A correlation of its incidence with dietary fat, Am.J.Med.Sci., 220:421-430.

Swank, R. L., 1970, Multiple Sclerosis: Twenty years on low fat diet, Arch.Neurol., 23:460-474.

Tamblyn, C. H., Swank, R. L., Seaman, G. V. F., and Zukoski, C. F.iv, 1980, Red cell electrophoretic mobility test for early diagnosis of multiple sclerosis, Neurol.Res., 2:69-83.

Thompson, R. H. S., 1966, A biochemical approach to the problem of multiple sclerosis, Proc.Roy.Soc.Med., 59:269-276.

Thompson, R. H. S., 1973, Fatty acid metabolism in multiple sclerosis, Biochem.Soc.Symp., 35: 103-111.

Tingley, A. H., 1956, Human brain lipids at various ages in relation to myelination, J.Ment.Sci.London, 102:851-855.

Turnell, R. W., Clark, L. H., and Burton, A. F., 1973, Studies on the mechanism of corticosteroid-induced lymphocytolysis, Cancer Res., 33:203-212.

Vost, A., Wolochow, A., and Howell, D. A., 1964, Incidence of infarcts of the brain in heart disease, J.Path.Bact., 88:463-470.

Waksman, B. H., 1959, Experimental Allergic Encephalomyelitis and the "Auto Allergic" diseases, Int.Arch.Allerg. and Appl.Immunol., Suppl. 14:1-87.

Yakovlev, P. I., and Lecours, A. R., 1967, The myelogenetic cycles of regional maturation in the brain, in: "Regional Development of the Brain", A. Minkowski, ed., Blackwell Scientific Publishers, Oxford and Edinburgh, 3-70.

Zollinger, H. U., 1941, Groszellig-granulomatöse Lymphangitis cerebri (Morbus Boeck) unter dem Bilde einer multiplen Sklerose verlaufend, Virchows Arch., 307:597-615.

PLASTICITY OF CONDUCTION PROCESSES IN DEMYELINATED NERVE FIBERS: IMPLICATIONS FOR THERAPY IN MULTIPLE SCLEROSIS

Charles L. Schauf

Departments of Physiology and Neurology
Rush University
Chicago, Illinois 60612

INTRODUCTION

Until fairly recently it had always been assumed that the complete loss of myelin over one or more internodes must necessarily produce an irreversible block of impulse conduction in myelinated nerve fibers. This view is a consequence of experimental observations showing that the essential ingredients for generation of nervous activity - voltage dependent channels selective for Na^+ ions - are normally only present at the nodes of Ranvier in mammalian and non-mammalian myelinated nerve fibers. Thus, the exposure of the internodal axon membrane which follows demyelination reveals a structure incapable of producing action potentials. A second obstacle to conduction following demyelination arises from the fact that, even if the internode were in principle excitable, the physiological function of the myelin sheath is to reduce the capacitance, and thus the amount of current necessary to charge the internodal membrane. Removal of this material, in the absence of any other changes, simply represents too great an electrical "load" on the excitation process to allow the continuation of conduction from the last normal node to the distal demyelinated segment(s).

In contrast to myelinated nerve fibers in which the impulse effectively "jumps" from one node to the next, in many mammalian fibers or nerve trunks (e.g. the vagus), as well as in all invertebrate species, impulse conduction occurs without an insulating myelin sheath. In this case, the machinery of excitation is distributed more or less uniformly along the fiber's length. At each point along its surface, the local nerve impulse provides the large current flow necessary to excite the immediately adjacent region of the fiber. The penalty such unmyelinated fiber systems pay for this mode of operation is greatly reduced conduction velocity, arising from the fact that each small area of membrane must be sequentially excited.

On the clinical side, everyone working with significant numbers of multiple sclerosis (MS) patients is struck by the immense variability of neurological signs and symptoms in this disease, both from one patient to another, and at different periods in the course of a particular individual's illness. It is not unusual for an MS patient to develop new findings, or undergo a dramatic exacerbation of existing symptoms, over just a few days; and then to spontaneously recover within an equally brief period. Occasionally, one even encounters the so-called "benign" patient who has symptoms so mild as to appear normal; and in many of these cases incidental autopsies have revealed widespread demyelination in functionally significant areas of the central nervous system (CNS), such as the optic nerve or dorsal column.

Currently, remyelination within the mammalian CNS, if it occurs at all, is considered to be a slow, very limited process (Prineas and Connell, 1979). Thus, it is necessary to reconcile a morphological situation which should not sustain impulse conduction, with the physiological fact that many nerve fibers may indeed be functional in the MS patient! The resolution of this apparent paradox lies in the recent observation that, at least in experimentally demyelinated nerve fibers, the machinery of impulse generation is sufficiently plastic to allow demyelinated regions of a nerve to behave as if they had never been myelinated in the first place - that is, to conduct impulses continuously over their entire length. If such a process also occurs in man, then it represents not only a key to unraveling the enormous variability that is seen clinically in MS, but also may point out a new direction for therapeutic investigations.

It is the aim of this Chapter to describe this axonal recovery process in some detail, and to discuss the pathophysiology of impulse conduction under such modified conditions. However, before embarking on these topics, it is necessary to briefly review the biophysical basis for impulse generation and conduction in nerve, and show how this becomes interrupted following demyelination.

101 CHANNELS IN MEMBRANES: THE BASIS OF EXCITABILITY

Perhaps the most fundamental biophysical advance of the last decade was the conclusive demonstration that there exist in nerve membranes quasi-permanent macromolecular pathways through which ions can move at rates exceeding 10^7 ions/second. These pathways have been termed channels because, at least during the time that they exist in a conducting ("open") configuration, the permeant ions move relative to the channel. This contrasts with carrier systems (such as ion "pumps") in which only a few ions are translocated each time the transport molecule moves across the membrane. Transmembrane ion channels come in all shapes and sizes, exhibit a wide range of physical behavior in different tissues, and in general may be regarded as responsible for all aspects of transmission and information processing in the nervous system.

In nerve fibers (as opposed to the cell bodies from which they arise) there exist only two sets of ion-selective channels, each operating quite independently of the other. One pathway permits only Na^+ ions to pass through, while the other is selective for K^+ ions. For the purposes of understanding the basic properties of the nerve impulse it is acceptable to consider the membrane to be impermeable to other cations, or to anions such as Cl^-. Each channel (Na^+ or K^+) behaves as if the entry of ions into it were controlled by a kind of "gate" whose state (open or closed) varies with changes in the voltage across the membrane. By using the term "gate" we do not mean that there must be a distinct structure controlling access to the interior. Most likely, the channel itself is only present in a conducting configuration some small fraction of the time, and it is this lifetime which depends on voltage. At any particular time each channel is either fully conducting or nonconducting, so that the membrane permeability is proportional to the fraction of total channels that are conducting. In the resting stage of a nerve fiber a small fraction of the K^+ gates are open, while essentially all the Na^+ gates are closed. Since the internal K^+ concentration is high, more K^+ ions tend to leave the cell than enter, and a (resting) potential develops such that the interior of the cell is negatively charged relative to the exterior (approximately - 70mv in most cells).

Suppose the membrane potential is suddenly displaced from its resting potential to a more depolarized value, and is then held there by appropriate electronic circuitry (this experimental procedure is termed a "voltage clamp"). By recording the transmembrane current, one can examine the way in which the Na^+ and K^+ gates respond to such changes (Figure 1). Here the number of conducting Na^+ and K^+ channels is shown as a function of time. When the membrane is slightly depolarized a few of the Na^+ channels open very quickly, and then spontaneously close despite the fact that the membrane is still depolarized (a process which is termed Na^+ inactivation). For larger depolarizations, the number of channels which open and subsequently inactivate, as well as the speed with which both processes occur, is increased. Finally, with sufficiently large depolarizations all the available Na^+ channels become conducting, and any further depolarization has little additional effect. Thus, the Na^+ permeability is a sigmoid function of membrane potential.

The potassium gates behave differently. Again progressively more open with increasing depolarization (Figure 1); however, not only is the time required for them to open much longer, but the K^+

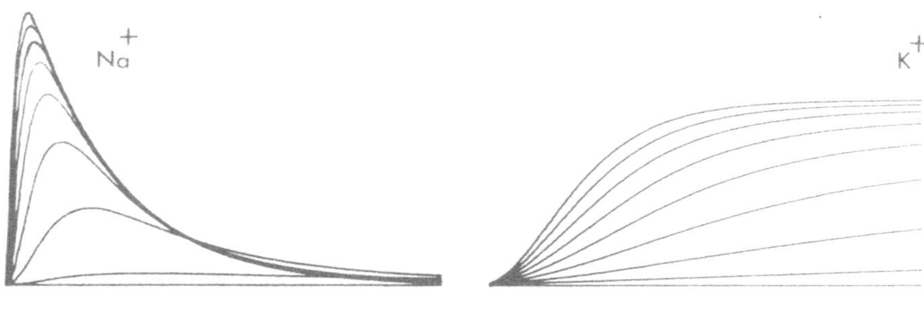

Fig. 1. Sodium and potassium permeability as a function of time after a sudden shift in membrane potential. The curves shown were calculated for depolarizations of 20-160 mV (in 20 mV increments). The na$^+$ and K$^+$ permeabilities can be assumed to equal the number of open channels (which has been plotted in arbitrary units).

gates remain open as long as the membrane is depolarized, shutting only when the membrane potential is restored to its initial state. They do not in general inactivate. The time course of K$^+$ activation is of the same order as the time course of Na$^+$ inactivation, and both are about 10 times slower than the process of Na$^+$ activation. This timing turns out to be critical in the production of a nerve impulse.

Figure 1 showed how the gates behave when we artifically depolarize the membrane and hold it there. Suppose we do not control the membrane potential, but permit it to react spontaneously to changes in the relative permeability of the membrane to Na$^+$ and K$^+$?

If we apply a pair of electrodes to a nerve fiber, and then pass an appropriate current between them, there will be a slight depolarization of the membrane in the vicinity of the cathodal electrode. This will cause a small fraction of the Na$^+$ gates to open, while little or no change occurs in the state of the K$^+$ gates simply because they react more slowly. A net inward movement of positive charge occur through the conducting Na$^+$ channels, while a net outward movement of positive charge takes place through those K$^+$ channels conducting at rest. If sufficiently few Na$^+$ channels open, the net rate of Na$^+$ entry will be less than the net rate of K$^+$ loss from the cell, and upon termination of the external current the cell's voltage will return to its resting level.

As the external current is increased, more Na$^+$ gates open and the net entry of Na$^+$ thus becomes progressively larger. For depolarizations of 15-20 mV the net inward movement of Na$^+$ begins to exceed the net outward movement of K^+. Then, even if the stimulating current is removed, the net flow of positive charge is into the cell, and this results in a further depolarization. This depolarization opens more Na^+ gates; the imbalance between Na^+ entry and K^+ exit is increased, leading to a larger depolarization; and another increment in the number of open Na^+ gates is produced. The cycle is regenerative and, in less than a millisecond, all the available Na^+ gates snap open. There are sufficient Na^+ channels so that the axon membrane is now more permeable to Na^+ than K^+, and thus the voltage must approach the equilibrium potential for Na^+ (reversed in polarity from the resting state since Na^+ is at high concentration externally). We have just described the phenomena of threshold. Either too few Na^+ channels open to overcome the tendency of the K^+ channels to repolarize the membrane, in which case the nerve repolarizes when the external stimulus is removed; or enough Na^+ channels become conducting to trigger the regenerative opening of the remainder, and a transition of the membrane potential to near the sodium equilibrium potential.

If activation of Na^+ channels were all that occurred, then a nerve fiber would remain depolarized indefinitely. However, once nearly all the Na^+ gates have opened, two additional factors come into play. In the first place, as mentioned earlier, Na^+ channels do not stay conducting. Once they have opened, they immediately begin to close again (inactivation). Simply by closing down Na^+ channels, the background leak of K^+ ions out of the nerve through those K^+ channels conducting at rest would eventually be able to repolarize the membrane. However, in addition, more potassium channels begin to activate at about the same time the Na^+ channels start to become inactivated, so that at any given instant during the falling phase of the action potential the membrane is relatively more K^+ permeable. Repolarization is thus accelerated, shortening the action potential. The potassium channels which are opened in response to a membrane depolarization are not inactivated, rather they simply shut down as the membrane repolarizes. The process of closing on repolarization takes some finite period of time, however, so that the potassium permeability remains slightly elevated even after the membrane has completely repolarized. The permeability changes during a typical action potential are illustrated in Figure 2.

It is a well-known classical phenomena that if one action potential is produced, a second cannot be elicited immediately following. The presence of such a refractory period is a consequence of Na^+ inactivation because the conformational state of the Na^+ channel following its opening and spontaneous inactivation is not identical with its resting state. There are at least three different configurations in which the Na^+ channel can exist - closed but activatable, conducting, and inactivated. Following activation into the conducting state, and subsequent passage into an inactive configuration, some period of time must pass before the channel finds itself in the closed but activatable state. Clearly if the Na^+ gates are inacti-

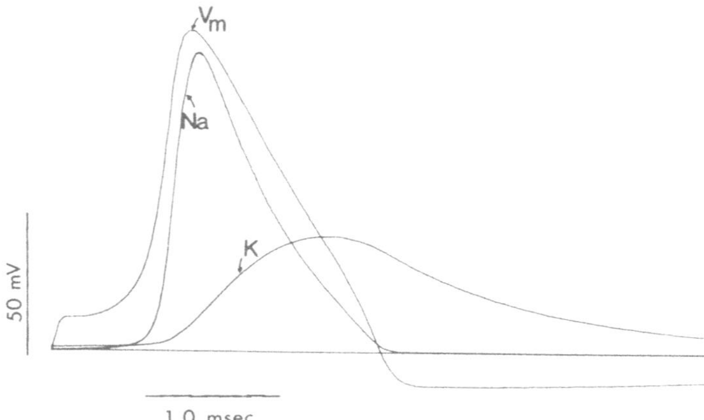

Fig. 2. Time course of membrane voltage (V_m), sodium permeability, and potassium permeability during a nerve impulse. Note that the nerve impulse is associated with a rapid increase in Na^+ permeability which reaches its maximum value near the peak of the impulse, and that the K^+ permeability remains elevated even after the membrane voltage has returned to near its resting level.

vated, and cannot open in response to a depolarization, there can be no action potential.

CONDUCTION OF THE NERVE IMPULSE

Up to this point we have been describing the production of a nerve impulse at a single point on a nerve fiber, while deliberately ignoring the problem of propagation. In order to understand the latter it is useful to initially consider the model myelinated nerve fiber shown in Figure 3 (top). Here areas at which the nerve fiber membrane is exposed (nodes of Ranvier - approximately 1 micron wide) are separated by long (millimeter or 1000 micron) stretches covered by multiple turns of Schwann or oligodendrocyte cell myelin. The details of this structure is discussed in standard texts (e.g. Gray's Anatomy) and reviews (Landon, 1982). It is sufficient to note that the multiple turns of membrane surrounding the internode act electrically to increase the transmembrane resistance (axon to external solution); and concommitantly to decrease its effective capacitance.

Suppose node A is producing an action potential at some particular instant of time. During the rising phase of the action potential at node A, Na^+ ions are entering the fiber - that is, there is an inward current. This current flows mainly down the axis cylinder (generally in both directions), with only a small amount leaking out

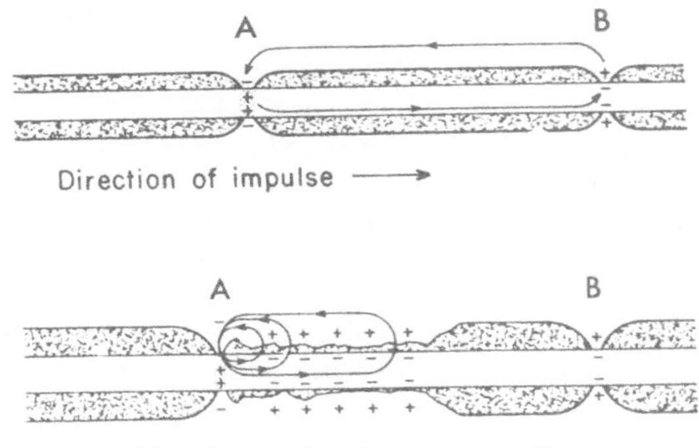

Fig. 3. An illustration of the conduction process in a normal myelinated nerve fiber, and in a "model" demyelinated nerve fiber. Note that the scale is greatly exaggerated. A typical node is about 1 micron wide, with an internodal distance perhaps 500-1000 times as great. The current flow is shown from an active node (node A) to the next most distal node (node B) in each case.

across the thick insulating layer of myelin. Upon reaching the next node, the positive charge neutralizes some of the excess negative charge on the inside of the membrane at node B. The now unbalanced positive charges on the outside of node B flow back to node A, completing the electrical circuit. The result of this process is a depolarization of node B to threshold, and the initiation of an action potential at B. The excitation sequence is then repeated. The action potential at node B again involves an inward movement of Na^+ (inward current), which spreads in both directions along the fiber, stimulating the next node; and so on for each successively more distal node of Ranvier. (The portion of current flowing backward from node B to node A can be neglected simply because the proximal node, having just produced an action potential, is refractory.) The action potential thus appears to "jump" from node to node down the length of the fiber, a process termed "saltatory" conduction.

What determines how fast such a nerve fiber conducts impulses? The key to this issue is that any membrane subjected to a step change in current does not change its potential instantaneously. Rather, the response to an applied current resembles that shown in Figure 4A. On increasing the current from one level to another, the voltage increases with a definite time lag. This delay arises because the membrane has a kind of electrical inertia (called "capacitance") which severely limits the rate at which voltages can change. Suppose we consider a nerve fiber subjected to a series of increasing current steps. The steady state value of the membrane potential will always be proportional to the applied current (Ohm's law), but in each case the voltage will increase much more slowly than the current. If the line in Figure 4A represents the threshold for action potential initiation then we can observe the following: (a) there are some currents too small to attain threshold, and (b) for two currents both giving steady-state voltages above threshold, the larger the current the briefer the period of time it must flow to drive the nerve to threshold. This forms the basis for what was termed the strength-duration relationship in the classical neurophysiology literature.

The considerations in the preceeding paragraph can be applied to the process of conduction in a myelinated nerve fiber. In flowing distally from one node to the next, some current leaks across the myelin sheath. The amount leaking out, compared to the amount flowing longitudinally, depends on the ratio of the membrane resistance to the internal resistance of the fiber (the larger the diameter the smaller its longitudinal resistance). The current loss can be quantitatively described by noting how far distally it is necessary to go to lose about 2/3 of the initial longitudinal current - this is termed the space constant. The larger the space constant, the less the attenuation of current from one node to the next, and the larger the remaining current. A larger exciting current means that a shorter time is required to reach threshold, permitting a higher overall conduction velocity. The space constant, and consequently the conduction velocity, is increased by increasing the membrane resistance of a nerve and/or by increasing the diameter of the fiber. Application of myelin is an extremely efficient way of increasing the membrane resistance, while simultaneously maintaining a small diameter. Conduction in a non-myelinated nerve fiber is similar in principle, except that the processes of excitation and current spread occur simultaneously in each small area of membrane. In a sense a non-myelinated nerve fiber may be visualized as a myelinated fiber with the internodal distance made infinitely short. Propagation occurs more slowly since there are in effect many more patches of membrane to excite (a myelinated fiber of 20 μ diameter has a conduction velocity approximately 10 times a as large as an unmyelinated fiber of the same diameter).

Under normal physiological conditions, because of the high resistance of the myelin sheath, the process of conduction in a myelinated nerve fiber involves the net delivery of current from an exciting proximal node to the next inactive node which is 5 to 7 times as great as necessary to drive the inactive node to threshold, and thus sustain conduction. This excess current is called the "safety factor". However, suppose a nerve fiber has one or more internodes in which a portion of the myelin sheath is absent or disrupted (Figure 3), such as occurs in multiple sclerosis. In this case relatively more current will be short-circuited between successive nodes. If we plot the potential of the inactive node as a function of time, we obtain the result illustrated in Figure 4B. The lower current means that more time is needed to excite the next node,

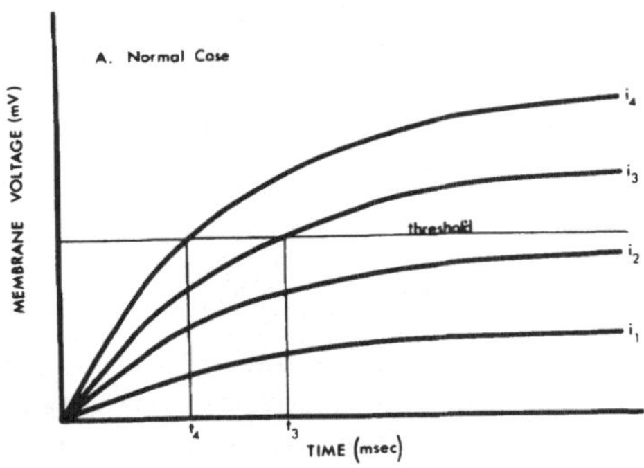

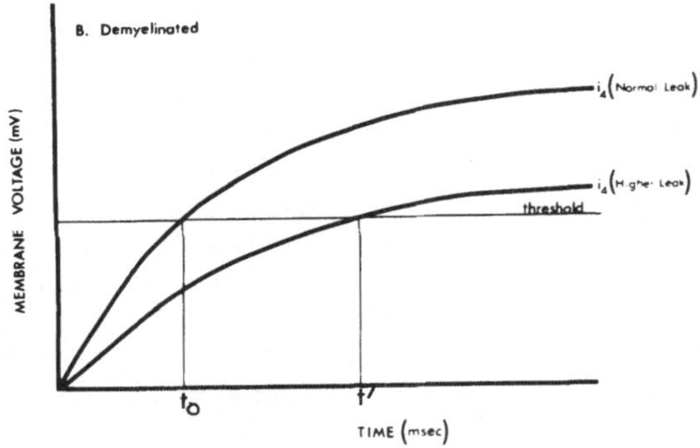

Fig. 4. Determinants of conduction velocity. In part A increasing currents (i_1, i_2, i_3, and i_4) are applied to a nerve. For the lowest currents, the final voltage is subthreshold. For the larger currents, the final voltage is above threshold, but for i_4 the time needed to reach threshold is less than that for the smaller current. For each curve, the time course is similar, and is determined by the effective capacitance. Since the conduction velocity is inversely proportional to the time needed to excite successive nodes, the larger the effective stimulating current (the safety factor), the greater the conduction velocity. The considerations of part A have been applied to the case of a demyelinated nerve fiber in part B. Here, the increased leakage of current in the demyelinated internode results in a decrease in the effective current reaching the distal node, increasing the time needed to reach threshold, and thus decreasing conduction velocity.

reducing the conduction velocity. If sufficient myelin is lost there may be so much leakage that not enough current will be available to depolarize the inactive node to threshold, and conduction will fail. What will be of particular interest to us in the remainder of this discussion are situations in which conduction is still possible; but where the total current leakage is sufficiently large so that the safety factor is reduced to a value near unity. Under such conditions the conduction process becomes very sensitive to minor changes in the local environment.

SPATIAL DISTRIBUTION OF EXCITABILITY

In considering conduction, a valid question that could be posed is why a demyelinated nerve fiber does not automatically revert to the same mode of conduction as is seen in normally nonmyelinated fibers - that is, a continuous propagation of the impulse along the demyelinated segments. As noted earlier, the answer is twofold. In the first place, even if the internodal membrane exposed as a result of demyelination contained excitable Na^+ channels, there would be a severe problem in initiating conduction from the last normal internode to the subsequent, demyelinated segments. The normal node of Ranvier is only 1-2 microns long. Thus, compared to the internode, the membrane area is quite small. Normally, the high resistance and low capacitance of the myelin sheath combine to dwarf the contribution of the internodal axon membrane to the overall properties of the nerve fiber. With demyelination, however, this situation is completely changed, and the properties of even a single exposed internode will dominate the behavior of the whole axon. The last normal node simply cannot deliver enough current to cause distal regions of the fiber to reach threshold, unless some other electrical characteristic of the fiber is also changed.

However, ignoring for a moment the problem of impulse initiation, a more serious difficulty is that myelinated nerve fibers only possess the machinery for excitation at the nodes of Ranvier. There are simply no Na^+ channels present under the myelin sheath. Support for this statement comes from two different kinds of experimental studies. In the first, neurotoxins that are known to specifically bind to Na^+ channels are radioactively labeled, and their uptake into whole nerve is measured on normally myelinated fibers; on fibers completely disrupted by vigorous homogenization; and on fibers immediately following demyelination caused by application of toxic substances such as lysolecithin. The result is that the exposure of an internodal membrane area perhaps 1000 times that of the node, results in no increase in toxin binding, implying a very sparse distribution of Na^+ channels along the normal internode (Ritchie and Rogart, 1977). Recently, we have been successful in examining the distribution of Na^+ channels more directly in individual nerve fibers using a radioactively labelled scorpion venom (Figure 5). When applied to longitudinal sections of normal myelinated nerve this Na^+ channel-specific toxin can be seen to be localized exclusively to the nodes of Ranvier. In contrast, label appears to be evenly distributed along nonmyelinated nerve fibers. Quantitatively, for rabbit myelinated nerve, the nodal Na^+ channel density is approximately 10,000/μ^2 (equivalent to a distance between channels of perhaps 100 A for a protein with a molecular weight of 250,000 daltons - very close indeed), while the upper limit on internodal channel density is perhaps 10/μ^2. By way of comparison, nonmyelinated fibers in the rabbit vagus nerve have 100 Na^+ channels/μ^2 distributed uniformly along their entire length (Ritchie and Rogart, 1977).

A second demonstration of the same point comes from electrical measurements. Here, single myelinated fibers are isolated and placed in an apparatus in which the currents flowing across a small length (50-100 μ) of the fiber can be accurately measured. If the area under study contains a node, large inward Na^+ currents are seen (Figure 6A), while in the internode the myelin sheath serves to block all current flow (not illustrated). Next, the paranodal myelin is acutely disrupted by the application of detergents such as lysolecithin; osmotic shock; or specific antisera. The magnitude of the Na^+ currents are unchanged (Figure 6B) even though one can demonstrate (by measuring total membrane capacitance) that there has been exposure of a significant amount of paranodal axon membrane. If the same experiment were to be performed at some point along the internode, no indication of an inward Na^+ current would be seen. If there were Na^+ channels beyond the node, they should have been observed following these procedures (Chiu and Ritchie, 1980; Smith and Schauf, 1981b).

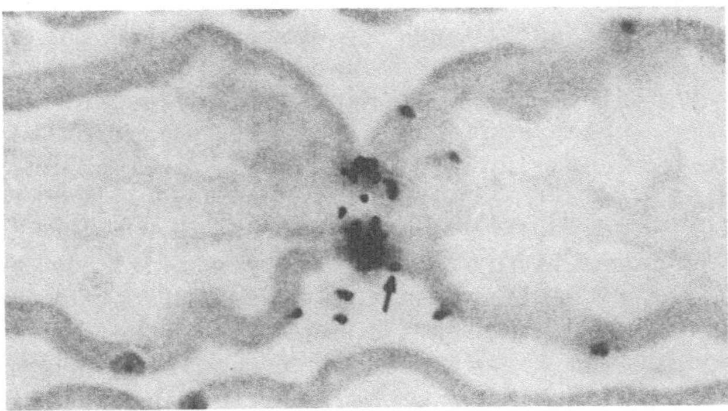

Fig. 5. Localization of Na^+ channels along a normal myelinated nerve fiber. Scorpion venom, a toxin which specifically binds to the Na^+ channel was labelled with I-125, and applied to a rat sciatic nerve. After incubation, the nerve was prepared for autoradiography. The dark grains represent venom molecules, and thus presumably Na^+ channels. Note the selective localization on the node of Ranvier and low background nonspecific binding.

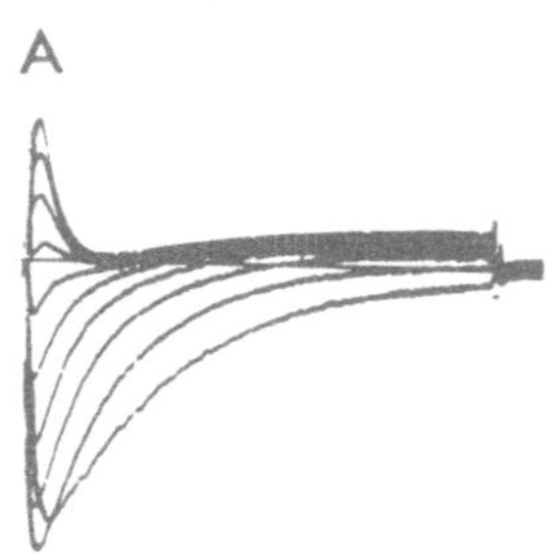

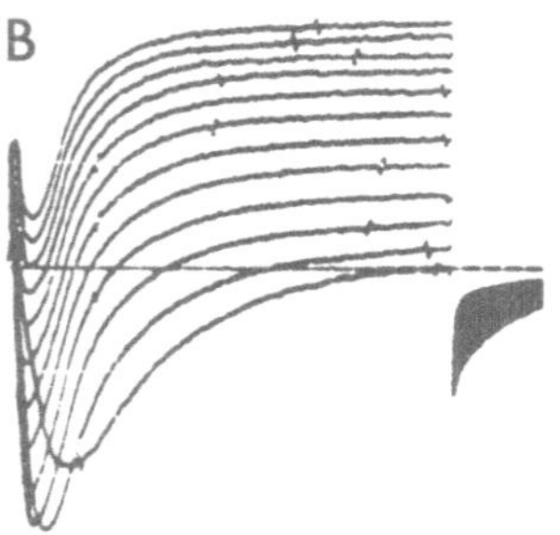

Fig. 6. Membrane currents at the node of Ranvier in a mammalian myelinated fiber before (left) and after (right) disruption of the paranodal myelin. In the normal situation only inward (at voltages below the Na^+ equilibrium potential) and outward (at voltages more positive than the Na^+ equilibrium potential) sodium currents are seen. There is a conspicuous absence of K^+ channels. Exposure of paranodal membrane does not increase the Na^+ current, but reveals the presence of K^+ channels which were previously beneath the myelin sheath, and thus electrically silent. (Data reproduced with permission from Smith and Schauf, 1981b).

Along with the selective localization of Na^+ channels at nodes of Ranvier, there seems to be a deliberate exclusion of K^+ channels from the same region. This is also evident from the records in Figure 6, where exposure of the paranodal membrane can be seen to cause a significant increase in the K^+ current. (In the internode, a similar disruption also reveals the presence of K^+ channels). The reason for such an exclusion is not known. Perhaps the high nodal Na^+ channel density, which must be deliberately controlled, simply precludes the presence of significant numbers of K^+ channels (whereas in nonmyelinated fibers with a much lower Na^+ channel density these two macromolecules have room to coexist). But why then have them in the internode which normally is electrically silent? Perhaps it is unnecessary to prevent their synthesis, given that other areas of the neuron clearly need such channels. In any case, the fact that such channels are present, and are exposed as a result of demyelination, has important implications for the therapeutic approaches to be discussed later.

RECOVERY OF EXCITABILITY FOLLOWING DEMYELINATION

Some 35 years ago Huxley and Stampfli (1949) first demonstrated the existence of saltatory conduction in myelinated nerve by recording (extracellularly) the currents flowing longitudinally along single nerve fibers. Recently a more refined, but essentially similar technique has been used to reveal the changes in conduction that follow demyelination in animal model systems (Rasminsky and Sears, 1972; Bostock and Sears, 1978, Smith, Bostock, and Hall, 1982). In this procedure, single rat motor fibers are selectively stimulated in the tail, while the longitudinal currents flowing in an undissected spinal root are recorded using very closely spaced electrodes. As the recording electrodes are moved along the roots, the differences between the longitudinal currents at successive electrode positions can be used to determine the sites of inward membrane current, corresponding to the location of voltage dependent Na^+ channels. In normal fibers, as expected, such sites of inward current are restricted to the usual internodal spacing (Figure 7). The experimental results are best presented visually as contour maps (see Bostock and Sears, 1978) in which the sites of inward current (solid concentric circles) and outward current (broken lines) are indicated as a dual function of distance and time. In this way both the distances between successive nodes (abcissa), as well as the time taken for the impulse to pass from node to the next (ordinate), can be readily identified. In the particular fiber shown in Figure 7 the conduction velocity was about 40 m/sec and the internodal distance approximately 1 mm.

Nerves can be experimentally demyelinated by a variety of techniques, but the use of the detergent lysophosphatidyl choline (lysolecithin - LPC) to disrupt the central and/or peripheral myelin

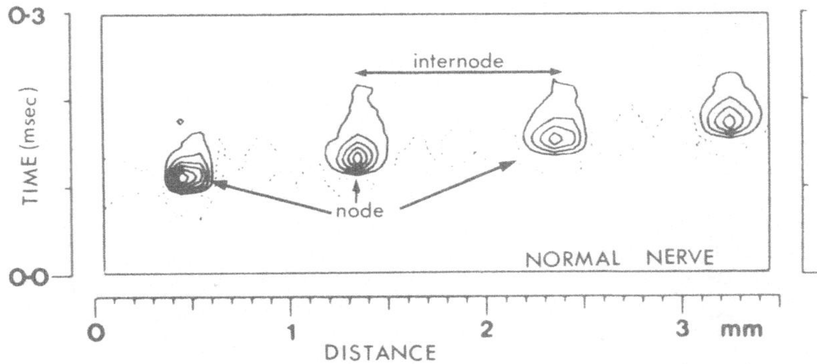

Fig. 7. The distribution of inward and outward currents in normal rat myelinated nerve. Inward (Na^+) currents are indicated by solid lines, and outward (K^+) currents by broken lines. The number of concentric circles is proportional to the density of current at a given point. The distance along the fiber is plotted on the x-axis, and the time on the y-axis. (Data reporduced with permission from Smith, Bostock and Hall 1982. The details of the experimental procedures used to generate such contour maps are described in Bostock and Sears, 1978).

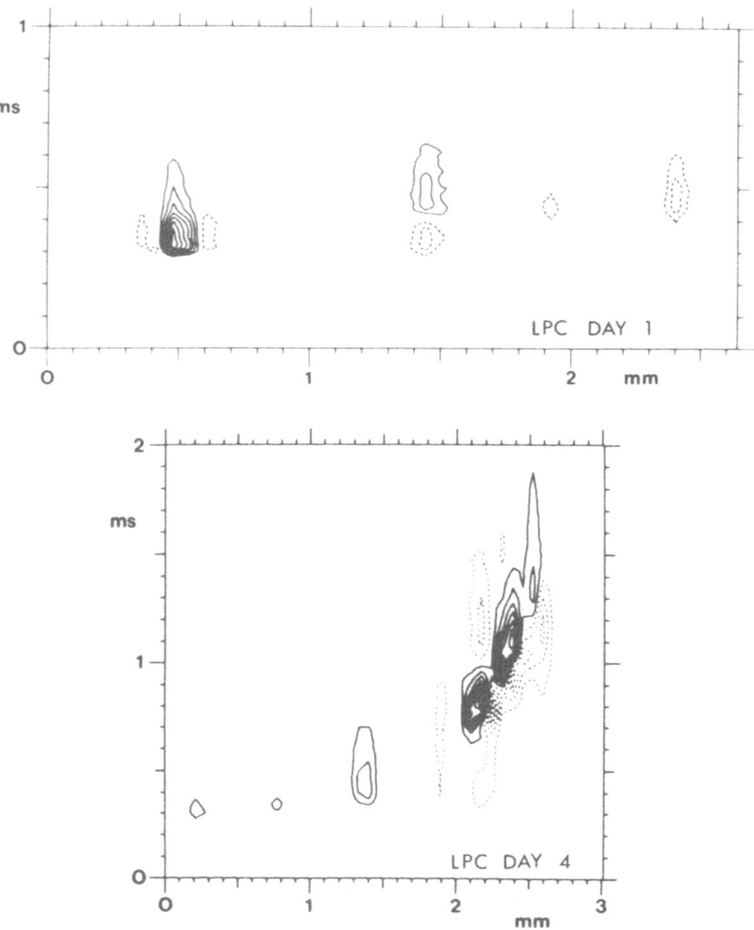

Fig. 8. Maps of current distribution similar to those shown in Figure 7 following the demyelination produced by lysophosphatidylcholine (LPC). The data in A were obtained at day 1 and illustrate the acute block of conduction, while those in B were from a lesion 4 days post LPC injection and show the initial appearance of sites of inward current in the newly exposed internode. Data reproduced with permission from Smith, Bostock, and Hall (1982).

sheath has proved to be particularly useful in examining the recovery processes which occur following demyelination because the events occurring in all the affected fibers in this lesion are well synchronized (Smith, Bostock, and Hall, 1982). As with other agents, rat nerves immediately following LPC treatment show segmental demyelination with a concommitant failure of impulse conduction (Figure 8A). There is no sign of any inward Na^+ current in the internode.

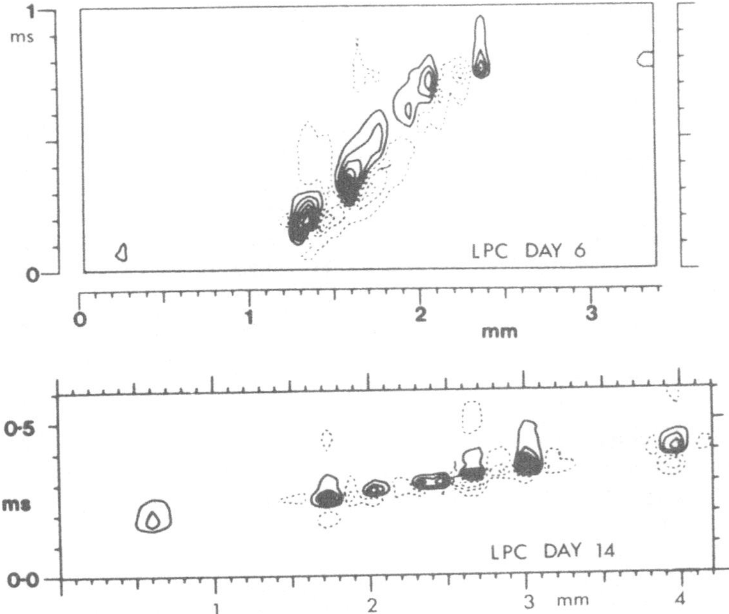

Fig. 8. (Cont'd). At 6 days, the contour map in part C shows that propagation occurs across the entire internode at low velocity, with normal conduction resuming beyond. However, within the internode, there is an uneven distribution of inward current. At 14 days following LPC injection, the countour map suggests that remyelination has occurred.

Remyelination begins only after 7-8 days following LPC treatment. However, as soon as 4 days post-demyelination, strong inward currents, not previously detectible, can be seen arising from axon membrane previously covered with myelin (Figure 8B). At this time the currents are usually not sufficiently large to sustain propagation across the entire internode. By 5-6 days, however, the recovery process has progressed to the point that conduction can occur over the length of the entire demyelinated internode, with normal conduction continuing more distally (Figure 8C). However, in the case of LPC treatment, internodal conduction does not become truly continuous, but remains "microsaltatory" with new, strong foci of inward current located some 100μ apart throughout the length of the internode. SInce no morphological evidence of remyelination is seen as this time, it must be the case that internodal Na^+ channels are nonuniformly distributed in the LPC lesion, perhaps due to a cellular control of the insertion sites. The distance between what have been termed "ϕ-nodes" (Smith, Bostock, and Hall, 1982) increases over the next few days, and eventually corresponds to the length of the new, shorter internodes in the remyelinated fibers (Figure 8D). It has thus been suggested that the axon, by localizing its Na^+ channels, may in fact be "instructing" the remyelinating glial cells where it is to place new myelin.

A somewhat different recovery process is seen following diptheria toxin induced demyelination (Bostock and Sears, 1978). As with lysolecithin, within a few hours following demyelination induced by the injection of diphtheria toxin into the cauda equina, the internodal conduction time is first increased (corresponding to a decrease in local conduction velocity to about 5-6 m/sec), and slightly later conduction is blocked, with no evidence of any inward Na^+ current beyond the last active node. Again, if one waits for a period of 4-6 days following the initial treatment with diphtheria toxin, recovery of conduction occurs.

In the earliest stages of this recovery, it is possible to resolve net inward Na^+ currents which extend beyond the last normal node into the previously inexcitable internodal region, but the impulse fails to propagate completely across the demyelinated internode. A day or two later, net inward currents can be detected throughout the internode, with local internodal conduction velocities that are comparable to those seen in normally nonmyelinated fibers of similar diameter (e.g. 102 n/sec in a fiber initially conducting at 40 m/sec). What is significant about the response to diphtheria toxin is that the countour maps show no sign of "hot spots", even at high spatial resolution. Thus, there is no evidence for an uneven distribution of inward currents along the internode (Figure 9) such

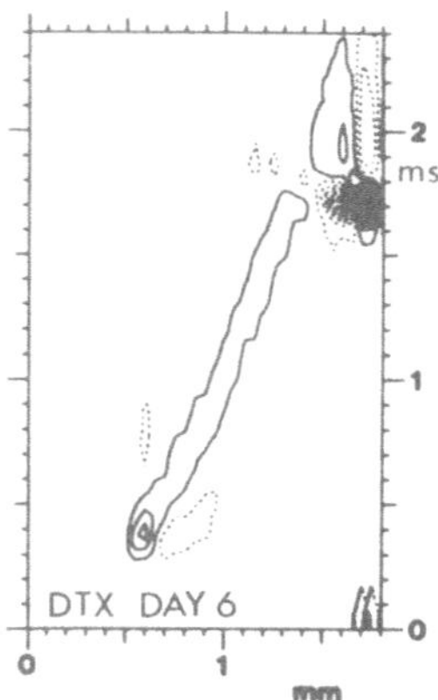

Fig. 9. Contour map in a rat myelinated nerve fiber 6 days after injection of diphtheria toxin. Again conduction throughout the internode is occuring, but now there is a relatively uniform inward current throughout the internode, suggesting a continuous distribution of Na^+ channels, such as might be seen in a normal, nonmyelinated nerve fiber. Data reproduced with permission from Smith, Bostock, and Hall (1982).

as seen with LPC. Beyond the area of segmental demyelination, conduction again proceeds normally in a saltatory fashion at high velocity.

Given the absence of internodal Na^+ channels in normal myelinated fibers, it is clear that the recovery of conduction following LPC and diphtheria toxin induced demyelination can only be explained by postulating the insertion of Na^+ channels (either at specialized sites or more or less uniformly) into regions previously devoid of them. Such an observation is perhaps not surprising when one realizes that remyelinated axons are clearly capable of conducting impulses, and characteristically have much shorter internodes than do the parent fibers. Obviously the sites of excitation have changed. Nevertheless, interesting questions are raised by these experiments. Do the internodal Na^+ channels represent newly synthesized and inserted macromolecules, or is there a spread of channels from the adjacent nodal regions? The fact that with diphtheria toxin one can often find demyelinated stretches extending over more than one internode, while the position of the "old" node is not marked by a corresponding peak in inward current, suggests that some decrease in nodal density occurs following demyelination. However, other experiments on a variety of excitable membranes argue that Na^+ channels may be relatively firmly anchored to the cytoskeleton and are not free of diffuse in the place of the membrane (Stuhmer and Almers, 1982). Probably what is occurring in both lesions is that the internodal Na^+ channels are being newly inserted, while those at the old nodes gradually turn over, and are not replaced at anything like their original high density.

HOW IS THE INTERNODE EXCITED?

It was noted earlier that a consideration of the passive electrical properties of a myelinated nerve fiber permit one to conclude that, even if the internodal region contains Na^+ channels, and thus is intrinsically excitable, the nerve impulse must still fail to invade a demyelinated internode because of the large electrical load it represents. Nevertheless, conduction can occur in experimental lesions. What is the mechanism responsible? Probably the changes in the Na^+ channel distribution along a demyelinated internode which allow conduction to occur are accompanied by alterations in passive fiber properties that facilitate the spread of an impulse into the demyelinated region (Waxman, 1978; Waxman and Brill, 1978). The adaptations that are needed to permit passage of the impulse across this transition region are well understood theoretically, but there is little or no experimental evidence in favor of any particular mechanism.

In general, the electrical problem is that, because of the increased capacitance, the last normal node usually delivers insufficient current to depolarize the internodal membrane to threshold.

Two kinds of adaptations are theoretically possible. On the one hand, there may be an increase in the current delivered by the proximal, exciting node. This may involve an actual change in its properties or, alternatively, the internodal length between the last few proximal nodes prior to the demyelinated stretch of axon may be decreased, thus decreasing the amount of current attenuation and increasing the effective safety factor for conduction. There is some morphological evidence of such changes in experimental lesions. Another possibility is that the new internodal Na^+ channels may have slightly different properties which effectively lower their threshold for excitation. Further studies are needed.

CAN HUMAN FIBERS RECOVER THE ABILITY TO CONDUCT IMPULSES?

Experimentally demyelinated fibers clearly can recover the ability to conduct electrical impulses in the absence of any significant degree of remyelination. Obviously it is impossible to perform in man the kind of electrical recordings that are used to demonstrate such internodal excitability in the rat. Moreover, multiple sclerosis lacks a completely comparable animal equivalent and, given the different responses that can be seen experimentally with LPC versus diphtheria toxin, one cannot *a priori* say very much about the expected sequelae to demyelination in man. Nevertheless, there is strong circumstantial evidence that some sort of recovery is possible, at least in a subset of the MS patient population.

Perhaps the strongest evidence for this kind of long-term plasticity in multiple sclerosis arises from the disparity between clinical signs and symptoms and pathological findings in significant fiber tracts referred to previously, and from the persistence of slowed conduction (as manifest in evoked responses) in the face of apparently normal neurological findings. More generally, provided one believes that remyelination within the human CNS is indeed rare, the fact that many patients recover a major portion of lost sensory and motor function itself suggests that internodal conduction of the type described does occur.

It is also significant that the clinical signs and symptoms in MS are completely understandable in terms of the conduction defects observed in experimental demyelination. Complete segmental demyelination in the absence of internodal excitability must result in conduction block. However, in experimental lesions, both paranodal demyelination, and segmental demyelination combined with the initiation of continuous (in the case of DTX) or microsaltatory (with LPC) conduction, yield fibers with reduced safety factors. Both conditions manifest themselves in terms of a marked slowing of conduction. It is quite common in multiple sclerosis to observe increases in somatosensory or visual evoked response latencies of 10-40 msec, even in individuals who have apparently recovered from acute exacerbations (McDonald, 1976; Halliday et al., 1977). Clearly slowing of impulse conduction in fiber tracts will most impair those functions (e.g. vibration sensitivity; ability to track moving objects; etc.) which depend on the delivery of precisely timed impulses in separate neural pathways.

Another characteristic of experimentally demyelinated fibers is their inability to faithfully transmit trains of impulses at high frequencies. Although able to conduct widely spaced impulses, the presence of a reduced safety factor causes such fibers to have an increased sensitivity to the small elevations in external K^+ concentration that are invariably associated with repetitive firing, leading to block at higher frequencies. Clinically, MS patients exhibiting sensory loss often show a decrease in critical flicker fusion frequency and loss of vibration sensitivity at frequencies as low as 40 Hz (Sclabassi et al., 1974).

In general, both paranodally demyelinated and segmentally demyelinated, continuously conducting fibers show decreased safety factors for conduction - the site in the latter case being at the junction between the last normal node and the demyelinated, conducting internode. It has been shown quite unequivocally that conduction in low safety factor fibers must have an increased sensitivity to small changes in temperature with heating producing block in conducting fibers and, conversely, cooling restoring conduction to blocked fibers (Schauf and Davis, 1974 - see following section). Clinically, such behavior is the basis for the hot bath test in multiple sclerosis, and has also been seen in investigations of the pattern visual evoked response after exercise (Persson and Sachs, 1978). Finally, experimentally demyelinated nerve fibers often show an increase in mechanical irritability (Howe et al., 1977), and such behavior mimics the clinical presence of Lhermitte's sign (Lhermette et al., 1924), visual phosphenes (Davis, et al., 1976), and paresthesias.

This close correspondence between the clinical pathophysiology of MS and the behavior seen in experimental lesions is not likely to be fortuitous. While paranodal demyelination can produce all of the foregoing phenomena, it is probable that in MS, as well as in the majority of toxic and allergic lesions, a complete segmental demyelination is the rule. If this is true, then conduction lability is most likely to arise at the low safety factor site between the last normal node and a demyelinated, excitable internode since, once it becomes excited, the internode itself would be expected to conduct quite securely (as would remyelinated fibers - see Smith and Hall, 1980; Smith, Blakemore, and McDonald, 1980).

LABILITY OF CONDUCTION IN MULTIPLE SCLEROSIS

If one accepts the argument that multiple sclerosis, like the toxic demyelinating lesions that have been more throughly studied physiologically, involves a conduction process which is proceeding with substantially reduced safety factors, then much can be learned by considering the pathophysiology of this situation in more detail. The precise events that reduce the safety factor become largely irrelevant, because the ways in which conduction can be blocked or restored under such conditions turn out to be quite indpendent of the exact details of the process (Schauf and Davis, 1974).

As noted above, minor changes in body temperature (less than 1°C) can produce dramatic, reversible alterations in neurological signs and symptoms in MS, with heat causing a worsening (Nelson and McDowell, 1959; Namerow, 1968) and cooling an improvement (Watson, 1959). It has been shown both theoretically and experimentally that this clinical phenomenon is due to a direct effect of temperature on the action potential duration in low safety factor demyelinated nerve fibers (Schauf and Davis, 1974; Davis et al., 1976; Rasminsky, 1978). Borderline conducting axons block with small temperature increases, whereas just barely blocked fibers are restored to conduction by a temperature decrease. The main effect of a temperature increase is to shorten the action potential duration. The decreased action potential duration leads to a reduction in inward Na^+ current. In a nerve fiber where the safety factor is already severely decreased to a value near unity, further lowering of this inward current by temperature elevation produces conduction block. Conversely, a lowering of temperature increases the action potential duration, and can restore conduction in borderline blocked demyelinated nerve fibers, no matter where the low safety factor site is. These phenomena are illustrated in Figure 10, which also shows that the conduction velocity itself depends on temperature in a way opposite to the degree of conduction block.

The conduction velocity of an unblocked nerve increases with increasing temperature, reaches a maximum, and then decreases slightly just prior to block. As the safety factor is lowered, conduction fails at much lower temperatures. Given a population of fibers with varying, but low safety factors, a small increase in temperature would cause a decrease in the number of conducting fibers. A demyelinated axon thus exists in a delicate balance between conducting and blocked states. Factors other than temperature which also act to slightly alter the safety factor can have dramatic effects, while causing little or no change in normal fibers. In fact, Schauf and Davis (1974) suggested that a practical symptomatic therapy for MS could be attained by using agents that could either lower threshold, and/or prolong the action potential by mimicking the effects of a temperature decrease - i.e. slowing Na^+ inactivation and/or inhibiting K^+ activation (Figure 11). An early example of

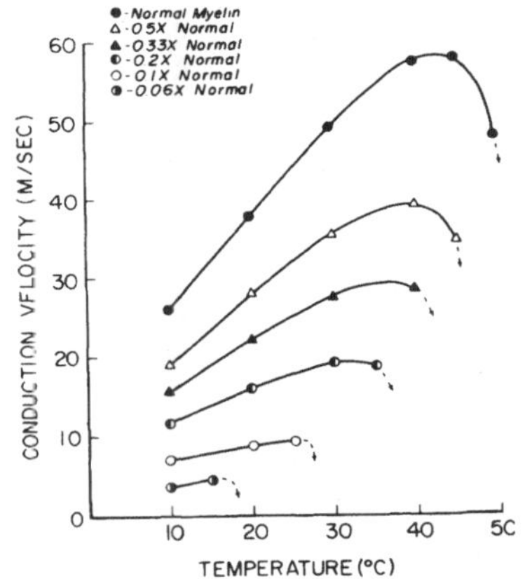

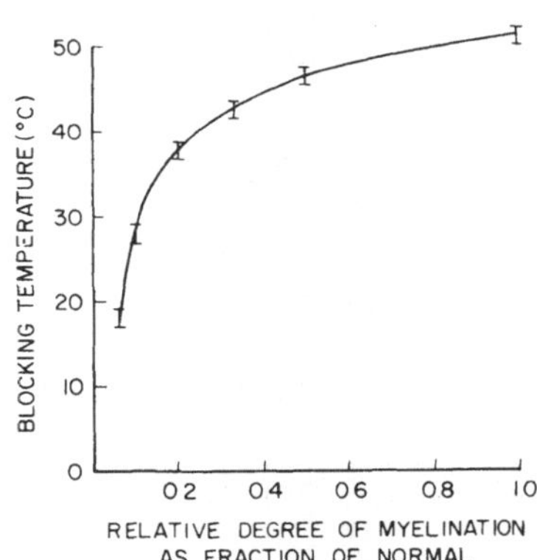

Fig. 10. The origin of the temperature sensitivity of conduction block. For a normal fiber the conduction velocity increases with temperature until conduction is blocked. In demyelinated fibers conduction velocity still increases with temperature, but the actual temperature at which conduction is blocked falls dramatically. This is illustrated more directly in B, where blocking temperature is plotted as a function of myelination. Data reproduced with permission from Schauf and Davis (1974).

this approach was the acute improvement in neurologic function in MS patients obtained by reducing plasma calcium levels (Davis et al., 1970). In later studies, Bostock et al., (1978) confirmed another prediction using scorpion venom, an agent known to slow Na^+ inactivation. In a rat spinal root preparation in which demyelination was produced by local injection of diptheria toxin, they were able to raise the blocking temperature of single demyelinated fibers by 4-12°C.

Reduction of the potassium permeability is also expected to improve conduction by increasing the blocking temperature, and Schauf and Davis (1974) first suggested that such drugs might prove useful. Although in normal fibers the potassium conductance plays little or no significant role in controlling action potential duration, internodal K^+ channels are exposed on demyelination. Both Pencek et al., (1978) and Sherratt et al., (1980) have demonstrated that 4-aminopyridine, a compound that inhibits the delayed potassium conductance at low concentration, facilitates conduction in blocked demyelinated nerve. Smith and Schauf (1981) recently reported a similar restoration of conduction by gallamine.

These ideas offer hope for the development of a symptomatic therapy. More work is needed on the pharmacological modification of

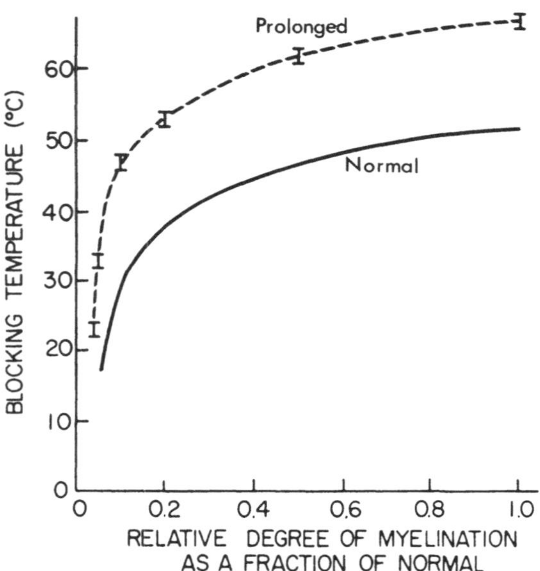

Fig. 11. The effects of prolonging the action potential (dashed curve) on the blocking temperature of a demyelinated nerve fiber. For any particular degree of myelin loss, slowing Na^+ inactivation or, alternatively, blocking K^+ channels, results in an increased blocking temperature. Data reproduced with permission from Schauf and Davis, 1974.

ion channels, and in careful drug studies in patients. Many of the drugs capable of restoring conduction in experimentally demyelinated fibers may have side effects that render them impractical for therapeutic use in man. In particular, agents that prolong action potentials in nerve fibers may have similar effects at nerve terminals, leading to increases in transmitter release. Such actions might have to be selectively blocked to eliminate side effects, unless some new level of specificity is discovered.

DO INTRINSIC INHIBITORS OF CONDUCTION EXIST IN MAN?

Given that many fibers in MS patients appear to be conducting with low safety factors, and are thus thermolabile, it is reasonable to inquire whether some intrinsic modulation of conduction other than temperature plays any role in the generation of transient clinical signs and symptoms in MS, without there being any substantial change in the degree of myelination itself.

Some years ago, we provided evidence that one or more factors capable of inhibiting the ventral root response in isolated spinal cord segments subjected to supramaximal dorsal root stimulation was present in the serum of MS patients. The activity of such "neuroelectric blocking factors" appeared closely correlated with variations in neurological signs and symptoms, both between different individuals and longitudinally within particular patients. Such blocking activity was not present in normal serum, and was eliminated or preserved by experimental manipulations in a manner consistent with the presence in MS sera of neurally directed antibodies (Schauf et al., 1981; Schauf and Davis, 1981).

Unfortunately, although such blocking activity is quite reproducible, the interpretation in terms of the role such antibodies have in the production of signs and symptoms in MS remains less certain. Multiple sclerosis is not thought to be a disease in which there occur significant changes in synaptic efficacy, and therefore for extrinsic agents to alter neurological signs and symptoms seems to require that they directly affect axonal conduction. We have shown that sera with potent blocking activity on spinal cord transmission do not have any detectable effect on conduction in either normal or demyelinated nerve. Rather, such sera inhibit both spontaneous and evoked release of neurotransmitters from nerve terminals. In addition, we have demonstrated that plasmapheresis consistently decreases blocking activity of patient sera, but does not in general alter clinical signs and symptoms. We were also unable to observe significant blocking activity in the cerebrospinal fluid itself.

Why, then, are anti-synaptic antibodies observed peripherally? It may be that the production of such blocking factors is the result of a generalized defect in immune regulation, leading to the production of antibodies to a variety of "nonsense" antigens unrelated to the primary event(s) of the disease process. This would be consistent with the presence of a variety of antiviral antibodies in oligoclonal CSF of MS patients, and would be expected to result in a situation in which levels of neuroelectric blocking antibodies were closely correlated with disease activity. However, their occurrence would then be a peripheral manifestation of the disease, rather than a cause of clinical exacerbations. Peripheral deficits have been observed in MS in studies of curare sensitivity and single fiber electromyographic jitter, although these particular experiments could not distinguish an effect on conduction in distal nerve terminals from a transmission defect. The anti-synaptic activity we have characterized would provide a sufficient explanation of these phenomena.

CONCLUSIONS

In summary, studies of experimentally demyelinated lesions over the past several years have provided an attractive explanation for the extreme lability of neurological deficits in multiple sclerosis patients, and for the degree of variability seen between different individuals affected with this disease. According to this hypothesis, the critical variable is the extent to which human demyelinated nerve fibers can develop the ability to conduct impulses in the same fashion as normal nonmyelinated fibers - that is, continuously over the entire length of the lesion. Studies of the basic cellular processes which control the synthesis and insertion into the membrane of excitable ion channels, as well as their detailed properties, then become most relevant to the search for an ultimate therapeutic intervention in this disease.

Acknowledgements

The original experiments described in this Chapter were supported by National Multiple Sclerosis Society Research Grant 1313B3 and by the Morris Multiple Sclerosis Research Fund.

REFERENCES

Bostock, H., and Sears, T. A., 1978, The internodal axon membrane: electrical excitability and continuous conduction in segmental demyelination, J.Physiol.(London) 280:273-301.

Bostock, H., Sherratt, R. M., and Sears, T. A., 1978, Overcoming conduction failure in demyelinated nerve by prolonging action potentials, Nature 274:385-387.

Chiu, S. Y., and Ritchie, J. M., 1980, Potassium channels in nodal and internodal axonal membrane of mammalian myelinated fibers, Nature (London), 284:170-171.

Davis, F. A., Schauf, C. L., Reed, B. J., and Kesler, R. L., 1976, Experimental studies of the dependence of conduction in normal and demyelinated nerve on extrinsic factors, I. Temperature. J.Neurol.Neurosurg.Psychiatry, 39:441.

Davis, F. A., Becker, F. O., Michael, J. A., and Sorensen, E., 1970, Effect of intravenous sodium bicarbonate, disodium edetate (Na_2EDTA), and hyperventilation on visual and oculomotor signs in multiple sclerosis, J.Neurol.Neurosurg.and Psychiatry, 33: 723-732.

Davis, F., Bergen, D., Schauf, C. L., McDonald, I., and Deutsch, W., 1976, Visual phosphenes in multiple sclerosis, Neurology, 26: 1100-1102.

Halliday, A. M., McDonald, W. I., and Mushin, J., 1977, Visual evoked potentials in patients with demyelinating disease, in: "Visual Evoked Potentials in Man," J. E. Desmedt, ed., Claredon Press, Oxford.

Howe, J. F., Loeser, J. D., Calvin, W. H., 1977, Mechnosensitivity of dorsal root ganglia and chronically injured axons: a physiological basis for the radicular pain of nerve root compression, Pain 3.

Huxley, A. F., and Stampfli, R., 1949, Evidence for saltatory conduction in peripheral myelinated nerve fibers, J.Physiol (London), 108:315-339.

Landon, D. N., 1982, The structure of the nerve fiber, in: "Abnormal Nerves and Muscles as Impulse Generators," W. Culp and J. Ochoa, eds, Oxford University Press, pp. 27-53.

Lhermitte, J., Bollack, J., Nicholas, M., 1924, Les douleurs a type de decharge electrique a la flexion cephalique dans le sclerose en plaque, Rev.Neurol., 2:56-62.

McDonald, W. I., 1976, Pathophysiology of conduction in central nerve fibers, in: "New Developments in Visual Evoked Potentials in the Human Brain," J. E. Desmedt, eds., Oxford University Press.

Namerow, N. S., 1968, Circadian temperature rhythm and vision in multiple sclerosis, Neurology (Minneap) 18:417-422.

Nelson, D. A., and McDowell, F., 1959, The effects of induced hyperthermia on patients with multiple sclerosis, J.Neurol.-Neurosurg. and Psychiatry, 22:113-116.

Pencek, T. L., Schauf, C. L., Low, P. A., et al., 1980, Disruption of the perineurium in amphibian peripheral nerve: Morphology and physiology, Neurology, 30:593-599.

Persson, H. E., and Sachs, C., 1978, Provoked visual impairment in multiple sclerosis studied by visual evoked responses, Electro-encephalographic J., 44:664.

Prineas, J. S., and Connell, F., 1978, The fine structure of chronically active multiple sclerosis plaques, Neurology, 28:68.

Rasminsky, M., 1978, Physiology of conduction in demyelinated axons, in: "Physiology and Pathobiology of Axons," S. G. Waxman, ed., pp.361-367, Raven Press, New York.

Rasminsky, M., and Sears, T. A., 1972, Internodal conduction in undissected demyelinated nerve fibers, J.Physiol.(London), 227:323-250.

Ritchie, J. M., and Rogart, R. B., 1977, The density of sodium channels in mammalian myelinated nerve fibers and the nature of the axonal membrane under the myelin sheath, Proc.Nat.Acad.Sci., (USA), 74:211-215.

Sclabassi, R. J., Namerow, N. S., and Enns, N. F., 1974, Somatosensory response to stimulus trains in patients with multiple sclerosis, Electroencephalography and Clinical Neuro physiology, 37:23.

Schauf, C. L., and Davis, F. A., 1974, Impulse conduction in multiple sclerosis: A theoretical basis for modification by temperature and pharmacological agents, J. Neurol.Neurosurg.Pyshciatry, 37:152-161.

Schauf, C. L., and Davis, F. A., 1981, Neuroelectric blocking factors in multiple sclerosis: a perspective, in: "Demyelinating disease: Basic and Clinical Electrophysiology," S. G. Waxman and J. M. Ritchie, eds., Raven Press, New York, p.267.

Schauf, C. L., Pencek, T. L., Davis, F. A., and Rooney, M. W., 1981, Physiological basis for neurelectric blocking activity in multiple sclerosis, Neurology, 31:1337.

Sherratt, R. M., Bostock, H., and Sears, T. A., 1980, Effects of 4-aminopyridine on normal and demyelinated mammalian nerve fibers, Nature (London), 283:570-572.

Smith, K. J., and Hall, S. M., 1980, Nerve conduction during peripheral demyelination and remyelination, J.Neurol.Sci., 48:201-219.

Smith, K. J., and Schauf, C. L., 1981a, Effects of gallamine triethiodide on membrane currents in amphibian and mammalian peripheral nerve, J.Pharmacol.Exp.Ther., 217:719-726.

Smith, K. J., and Schauf, C. L., 1981b, Size-dependent variation of nodal properties in myelinated nerve fibers, Nature, 293:297.

Smith, K. J., Blakemore, W. F., and McDonald, W. I., 1980, Central remyelination restores secure conduction, Nature (London), 280:395-396.

Smith, K. J., Bostock, H., and Hall, S. M., 1982, Saltatory conduction precedes remyelination in axons demyelinated with lysophosphatidyl choline, J.Neurol.Sci., 54:13-31.

Stuhmer, W., and Almers, W., 1982, Photobleaching through glass micropipettes: sodium channels without lateral mobility in the sarcolemma of frog skeletal muscle, Proc.Nat.Acad.Sci., 79: 946-950.

Watson, C. W., 1959, Effect of lowering of body temperature on the symptoms and signs of multiple sclerosis, New England J.Med., 261:1253-1259.

Waxman, S. G., 1978, Prerequisites for conduction in demyelinated fibers, Neurology, 28:27-33.

Waxman, S. G., and Brill, M. H., 1978, Conduction through demyelinated plaques in multiple sclerosis, Computer simulations of facilitation by short internodes, J.Neurol Neurosurg. Psychiatry, 41:408-416.

RECENT STATUS OF EPIDEMIOLOGICAL STUDIES ON MULTIPLE SCLEROSIS IN ITALY

Claudio Mariani, *Enrico Granieri and Guglielmo Scarlato

2nd Neurological Clinic of the University of Milan
20122 Milano, Italy. *Neurological Clinic of the
University of Ferrara, 44100 Ferrara, Italy

INTRODUCTION

Multiple sclerosis (MS) is difficult to study by epidemiological methods, since it shows unfavorable characteristics for the epidemiologic approach.

In fact, its low frequency, long latency and long duration reduce the possibilities of the epidemiological approach, and, for these reasons, almost exclusively descriptive studies are used, above all prevalence surveys, which are, however, more suitable for quantifying the condition of the general public health than for answering etiological questions.

In contrast analytic studies, namely retrospective or case-control studies, and prospective or cohort studies, electively advisable to test the hypothesis resulting from descriptive surveys, clinical observations and laboratory studies, are either difficult or impossible to carry out. In fact, a MS case-control study ought to be devised for investigation into childhood and adolescence events, but the difficulties involved are obvious. Moreover, prospective studies are practically impossible to organize because it would be necessary to keep thousands of people under observation for many years.

Since low frequency prevents the conduction of door-to-door inquiries, MS cases can only be discovered from various medical sources. So, qualitative level, organization and accessibility to these sources are decisive factors in determining reliability and comparison of the estimates.

Given these characteristics of the disease, when the sources of study material have not undergone essential changes in time, the use of incidence is not very reliable as compared to prevalence. For these reasons, most authors have always used prevalence. However, it must be emphasized that incidence expresses the proportion of individuals belonging to a given population at risk who develop a disease in a definite time interval: as for MS, the incidence measures the frequency of additional new cases of the disease per 100,000 per year; thus, the incidence rates are important indicators of the temporal trend of MS and, therefore, besides being necessary for preventive measures, they can provide etiological clues.

In contrast, prevalence measures the proportion of individuals belonging to a specific population at risk who have the disease at a given point in time: for MS, at prevalence day. The prevalence rate, by assessing the frequency of all current cases of disease (existing cases and new cases) within a specific population, quantifies the current need and is particularly useful as an indicator for medical and social care to cope with the existing cases. Contrary to incidence, prevalence is likely to be biased by other geographic variables such as migratory flux within the population, accessibility to the neurological centers, level of assistance and consequent survival of the sick. These problems, more evident in a disease with particular characteristics such as MS, tend to make the prevalence unsuitable for comparing prevalence rates found in different areas and years.

Long latency and wide variability of the interval between onset and diagnosis cause an under estimation of the true frequency.

Finally, in MS a really specific and easily feasible diagnostic test does not exist. The diagnostic criteria of Allison and Millar (1954), McAlpine et al., (1965), Schumacher committee (1965), Rose et al., (1976) and McDonald and Halliday (1977) still used in epidemiology are all descriptive and classify the MS cases according to the probability of the diagnosis. In the processing of the data, the MS epidemiological studies consider only the cases in which the two more important clinical criteria are satisfied simultaneously: the temporal dissemination of the symptoms and the spatial dissemination of the signs. The above mentioned clinical diagnostic classification (probable cases of Allison and Millar, definite cases of Schumacher, clinically definite cases of McAlpine) satisfy all such criteria, and correspond to one another. However, the use of descriptive clinical criteria alone, without support of paraclinical tests and laboratory procedures, gives rise to subjective bias and so can introduce unquantifiable variables when accepting MS cases.

These characteristics of MS pose serious methodological problems. We feel that they were worth mentioning but they should not be a matter of scepticism. In fact, in spite of all these limits, the use of the epidemiological approach has brought to light many pathogenetic connections that might otherwise never have been revealed by other investigative methods. We should like to bring your attention to an essential point: To carry out epidemiological studies means to work with constant care in order to pick out each possible source of bias. A survey can be considered valid if it has been able to minimize the biases, since it is impossible to remove them all in the observational studies (Rosati, 1981).

Despite these limits, the epidemiological approach is still one of the eventual hopes for a better knowledge of the enigmatic nature of MS. About 250 descriptive studies on MS prevalence in different areas of the world have been performed in the last 30 years, which allows us to elaborate a pattern of the natural history of MS casting some light on etiological research.

At first, the current opinion was that MS was differently distributed throughout the world, and its frequency was a function of geographic latitude: the higher the latitude, the higher the MS prevalence rate (Alter, 1968; Kurland, 1970).

Soon afterwards, according to data from many other prevalence surveys, a division of the world into three parts for the distribution of MS was proposed by Kurtzke in 1975. Such distribution was confirmed by the same author in 1980, on the basis of 234 prevalence surveys. High frequency rate of 30 cases, or more, per 100,000 population is limited to northern and central Europe (especially in the United Kingdom where highest figures were estimated), southern Canada, and the United States north of 37° north latitude, Australia and New Zealand, and South-Africa. In other words, MS seems to involve Europe and those extra-European countries in which, throughout various historical periods, there were strong settlements of Europeans, and in particular the British colonies. According to some opinions, this pattern of distribution could indicate that MS originated in Europe, and was spread through the rest of the world by means of the strong migrations of northern Europeans. Such an inference points to the epidemic nature of the disease. Recently, several authors, prompted by two epidemic foci in Faroë Islands and in Iceland, are trying to test this hypothesis through the search of temporal and spatial clusters (Poskanzer et al., 1981).

Medium frequency regions with rates between 5 and 25 per 100,000 population comprise northern Scandinavia, the Mediterranean basin, the southern United States, most of Australia, and perhaps the middle part of South America.

Asia seems to be globally, or nearly, a low risk band, with the exception of western Siberia and Transcaucasia.

Apart from one white group of South Africa, all known parts of the African continent are low risk areas, even if it must be emphasized the the MS prevalence of 10, estimated in the African littoral, in Tunisia, raises the possibilities that Afro-Asiatic Mediterranean countries might share the MS prevalence of the northern regions of the Mediterranean.

That Mediterranean Europe is a medium frequency zone, is not at all certain. On one hand, when considering the Iberian peninsula, we have no descriptive study to refer to; but, if we set a value on the Spanish mortality rate of 1.0 per 100,000 established 20 years ago by Dominguez Carmona (1961), Spain ought to be included within the high risk zones. From Greece and the Balkans, data are still lacking: the only prevalence study carried out using only in-patients seen at the Neurological Clinic of Tessaloniki, offers, therefore, an underestimate rate. But a good study from Rijeka city, on the high Adriatic littoral, puts Yugoslavia into the high frequency zone.

However, the current opinion that Mediterranean countries are medium frequency areas for MS is to a great extent based on the results of about 30 Italian prevalence studies between 1962 and 1982. These surveys, reporting MS probable prevalence rates from 7.2 to 27.1 per 100,000, were undertaken on large populations, generally exceeding 300,000 inhabitants, that is to say, a methodological approach was used which in Italy involves some problems of organization and thus increasing the risk of bias.Surveys on large populations can be exposed to errors in estimates, especially when one works in a health organization like the Italian one, which is not very suitable for epidemiologic research. The figures of such studies in Italy cannot be compared to those found in northern and central European countries, which have a historically greater experience in the epidemiological field. Current medical organization in Italy shows several defects including poor accessibility to the sources of study material, low level of organization in many hospital archives, differences in systems of medical records, shortage of rehabilitation facilities for the handicapped, tendency of many patients to attend medical centers outside their health districts, and lack of up-to-date population data. These problems tend to underestimate the true prevalence, especially when large populations are studied.

Likewise, the reported crude prevalence rates, found in 19 Italian provinces, would indicate a decreasing frequency from north of Italy to southern regions (Table 1). However, this finding, too is open to criticism. Foremost, lower figures in southern Italy probably reflect major incompleteness in case-collection practices in that part of the Peninsula owing to the lower level of public health services. Secondly, it must be underlined that no prevalence rate was standardized from the study area's population for both age and sex. In Italy the composition of southern populations is quite different from that of northern populations, in which the proportion of individuals in the middle age groups is much higher. This diff-

Table 1. MS Prevalence Studies Carried Out in Italy Using Large Populations

AUTHORS		STUDY AREA	LATITUDE	POPULATION	PROBABLE MS PREVALENCE RATES PER 100,000	95% C.I.	STANDARDIZED PREVALENCE RATES PER 100,000
Marforio	(1980)	Sondrio	46°	174,327	24.1	17-33	-
Rossi	(1980)	Trento	46°	442,873	27.1	22-32	-
Marforio	(1980)	Varese	45°	790,046	24.2	21-28	-
Mamoli	(1980)	Bergamo	45°	891,153	15.8	13-19	-
Bortolon	(1982)	Vicenza	45°	727,296	32.2*	28-37	-
Caputo	(1979)	Novara	45°	507,394	19.5	16-24	-
Borri	(1976)	Torino	45°	2,287,016	17.1	15-19	-
Nardi	(1979)	Venezia	45°	831,657	18.6	16-22	-
Tavolato	(1974)	Padova	45°	760,649	16.0	13-19	17.8
Rosati	(1981)	Ferrara	44°	386,896	26.9	22-33	24.8
Macchi	(1962)	Parma	44°	389,279	11.6	8-15	-
Borri	(1976)	Firenze	43°	1,146,367	10.6	9-13	-
Borri	(1976)	Perugia	43°	552,136	17.0	14-21	-
Paci	(1980)	Terni	43°	229,034	17.9	13-24	-
Megna	(1980)	Bari	41°	1,351,288	15.5	13-18	-
Palma	(1981)	Avellino	40°	433,864	7.6**	5-11	-
Palma	(1981)	Napoli	40°	2,812,996	9.1**	8-10	-
Rosati	(1977)	Sassari	40°	397,891	16.6	13-21	17.6
Rosati	(1977)	Nuoro	40°	273,021	15.8	12-21	17.9
Rosati	(1977)	Cagliari	39°	802,888	9.5	8-12	10.2
Bramanti	(1978)	Messina	38°	673,791	12.6	10-15	-
Savettieri	(1978)	Palermo	38°	666,175	12.3	10-15	13.1
Reggio	(1979)	Catania	37°	938,273	7.2	6- 9	-

* Includes suspected and possible MS.

** Includes possible MS.

erence of the age composition of the Italian population clearly appears in Figure 1, which compares the Emilia Romagna population, northern Italy, with the Sardinian population, insular-southern Italy. The Emilian population structure is columnar, with a high proportion of individuals in the mid-age classes; in contrast the Sardinian population is pyramidal, with a lower proportion of individuals in middle age. The crude rates estimated in these two popula-

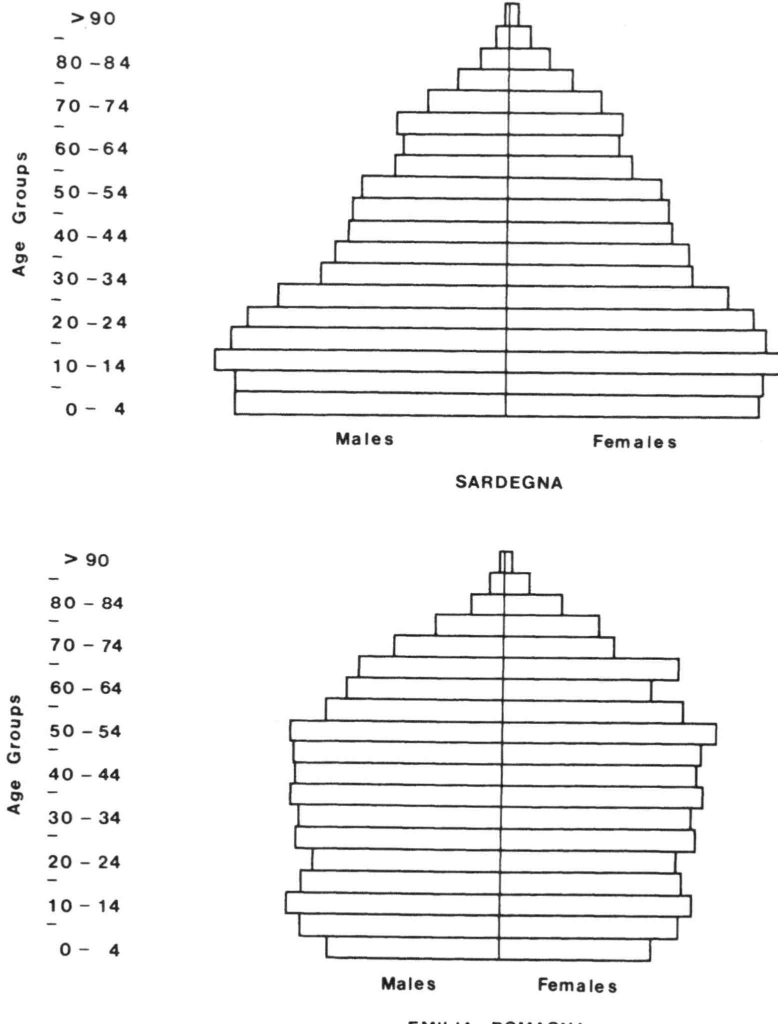

Fig. 1. Population structures according to age- and sex-composition in Sardinia and in Emilia Romagna in 1971 (Rosati and Granieri, 1982).

tions are not comparable. As we adjust rates versus the Italian population, the rates found in Emilia Romagna decrease, and those found in Sardinia rise considerably.

An indirect confirmation of the limits in comparing international data carried out using large populations is suggested by the results of a study of Dean et al., (1976) on the frequency of first admissions for MS to London hospitals among Italian immigrants. The survey seems to indicate that Italian immigrants have an analogous risk to that observed among the English. In fact, the number of first admissions found among the Italian community was like that expected for the British population. So, it is not very credible that Italians who have migrated to London increase their risk for MS: moreover, it would contradict most studies on migrants. This observation, on the contrary, stresses the possibility that the true MS frequency in Italy is much higher than reported by Italian studies on large populations.

Attempts, have, therefore, been made to verify the MS prevalence in Italy by using more appropriate methods to minimize the risk of biased data. Since it is not possible either to change rapidly the current health situation, or to increase the efficiency of the studies (and thus the precision of the rates), some authors have decided to concentrate all their efforts, including increase in staff and expenditure, in relatively small homogeneous communities (30,000 - 50,000 inhabitants). On this basis, Dean et al., (1979) first contested the inclusion of Italy within the medium-risk zone with his survey of Enna in Sicily, where a rate of 53 per 100,000 was recorded. Soon after, between 1979 and 1982, the results of other 7 intensive surveys were published, offering considerable support to Dean's study. All these studies estimated prevalence rates higher than 30 per 100,000 (Table 2), and are all concordant in implying that Italy is also a high-frequency zone, and in suggesting that the frequency does not decrease from North to South according to a gradient of latitude. Moreover, with regard to the importance of adjusting the rates versus the Italian population for both age and sex, we must take into consideration the rates estimated both in northern Italy (Copparo in Emilia Romagna) and in southern Italy (Barbagia and Ozieri in Sardinia). The standardized rates from Copparo decrease, and those from Barbagia and Ozieri increase considerably (Rosati et al., 1980a; Rosati et al., 1980b; Granieri and Rosati, 1982).

In conclusion, it is likely that accurate estimates of MS prevalence in Italy can be obtained only by intensive surveys of limited population groups. Surveys of this kind, although too few and exclusively based on prevalence estimation, do not permit definitive conclusions. However, they are comparable in indicating that the risk for MS is similar to that of most communities in central and northern Europe. In other words, all these findings, supporting the possibility that Italy is a high-risk area, seem to refute the "zonal

Table 2. MS Prevalence Studies Carried Out in Italy Using Limited Population Groups

AUTHORS	STUDY AREA	LATITUDE	POPULATION	PROBABLE MS PREVALENCE RATES PER 100,000	95% C.I.	STANDARDIZED PREVALENCE RATES PER 100,000
Rossi et al. (1980)	Bassa Valsugana	46°	24,570	44.8	22-80	-
Rosati et al. (1980a)	Copparo	44°	45,153	31.1	17-52	29.1
Dean (1981)	San Marino	43°	25,000	32.0	14-63	-
Rosati et al. (1980b)	Ozieri	40°	56,294	30,2	18-48	35.0
Granieri et al. (1982)	Barbagia	40°	51,611	40.7	25-62	48.5
Savettieri et al. (1981a)	Monreale	38°	25,403	43.0	22-77	-
Dean et al. (1981)	Agrigento	37°	49,979	32.0	18-52	-
Savettieri et al. (1981b)	Caltanissetta	37°	60,022	38.0	24-57	-
Dean et al. (1979)	Enna	37°	28,189	52.9	30-88	-

concept" for MS distribution in Europe and, therefore, the theory of the risk gradient with geographic latitude.

To confirm this hypothesis, many other similar investigations on homogeneous communities are urgently needed in Italy. In fact for two years now the Section of Neuroepidemiology of the Italian Society of Neurology has been concentrating most of its efforts to this end, having planned a series of descriptive and analytic studies in some Italian regions, which have different climatic situations and socio-economic conditions. These multicentric studies aim at ascertaining, once and for all, the true frequency of MS in Italy, and, at the same time, hope to discover factors, both genetic and environmental, that may play a role in the natural history of the disease in our country.

REFERENCES

Allison, R. S., and Millar, J. H. D., 1954, Prevalence and familial incidence of disseminated sclerosis: a report to the Northern Ireland Hospitals authority on the results of a three year survey, Ulster Med.J., 23:1.

Alter, M., 1968, Etiologic considerations based on the epidemiology of multiple sclerosis, Am.J.Epidem., 88:318.

Dean, G., McLoughlin, H., Brady, R., Adelstein, A. M., and Tallet-Williams, J., 1976, Multiple sclerosis among immigrants in Greater London, Br.Med.J., 2:861.

Dean, G., Grimaldi, G., Kelly, R., and Karhausen, L., 1979, Multiple sclerosis in southern Europe. I: Prevalence in Sicily in 1975, J.Epidem.,Commun.Health, 33:107.

Dean, G., Savettieri, G., Taibi, G., Morreale, S., and Karhausen, L., 1981, The prevalence of multiple sclerosis in Sicily. II: Agrigento city, J.Epidem.Commun.Health, 35:118.

Dominguez Carmona, M., 1961, Esclerosis en placas. Epidemiologia en Espana, Rev.Sanid,Hig.Pùbl., 35:113.

Kurland, L. T., 1970, The epidemiologic characteristics of multiple sclerosis, in: "Multiple Sclerosis and other Demyelinating Diseases," P. J. Vinken and G. W. Bruyn, eds., North Holland Publ.Co., Amsterdam.

Kurtzke J. F., 1975, A reassessment of the distribution of multiple sclerosis, Part one, Part two, Acta Neurol.Scand., 51:110.

Kurtzke J. F., 1980, Geographic distribution of multiple sclerosis: an update with special reference to Europe and the Mediterranean region, Acta Neurol.Scand., 62:65.

Granieri, E., and Rosati, G, 1982, Italy: A medium- or high-risk area for multiple sclerosis? An epidemiologic study in Barbagia, Sardinia, southern Italy, Neurology, 32:466.

McAlpine, D., Lumsden, C. E., and Acheson, E. D., 1972, "Multiple Sclerosis. A reappraisal," Churchill Livingstone, Edinburgh.

McDonald, W. I., and Halliday, A. M., 1977, Diagnosis and classifications of multiple sclerosis, Br.Med.Bull., 33:4.

Poskanzer, D. C., Walker, A. M., Prenney, L. B., and Sheridan, J. L., 1981, The etiology of multiple sclerosis: temporal-spatial clustering indicating two environmental exposures before onset, Neurology, 31:708.

Rosati, G., Granieri, E., Carreras, M., and Tola, R., 1980a, Multiple sclerosis in southern Europe: a prevalence study in the socio-sanitary district of Copparo, northern Italy, Acta Neurol. Scand., 62:244.

Rosati, G., Granieri, E., Carreras, M., Pinna, L., Tola, R., and Paolino, E., 1980b, Epidemiologia della sclerosi multipla in piccole communità dell'Italia Settentrionale ed Insulare, in: "Atti del Secondo Convegno Nazionale di Neuroepidemiologia," R. Boeri, and G. Filippini, eds., Associazione Promozione Ricerche Neurologiche, Milan, December 12-13, 1980.

Rosati, G., 1981, Epidemiologia della sclerosi multipla, in: "Atti XXII Congresso Nazionale della Società Italiana di Neurologia, Scalea, November, 5-7, 1981," T. Caraceni, ed., Crippa & Berger s.p.a., Milan.

Rosati, G., and Granieri, E., 1982, Ruolo degli studi descrittivi nella comprensione della eziologia della sclerosi multipla: analisi critica e proposte, in: "Atti del Terzo Convegno Nazionale di Neuroepdiemiologia," R. Boeri, and G. Filippini, eds., Associazione Promozione Ricerche Neurologiche, Milan, February 26-27, 1982.

Rose, A. S., Ellison, C. W., Myers, L. W., and Tourtellotte, W. W., 1976, Criteria for the clinical diagnosis of multiple sclerosis, Neurology, 26 (Suppl.):20.

Rossi, G., Ferrari, G., and Dalri, E., 1980, Sclerosi multipla nella provincia di Trento: ricerca epidemiologica, in: "Atti del Secondo Convegno Nazionale di Neuroepidemiologia," R. Boeri, and G. Filippini, eds., Associazione Promozione Ricerche Neurologiche, Milan, December 12-13, 1980.

Savettieri, G., Karhausen, L., and Dean, G., 1981a, The prevalence of multiple sclerosis in Sicily: Monreale city. J.Epidem.Commun. Health, 35:114.

Savettieri, G., Grimaldi, G., Giordano, D., Ventura, A., Karhausen, L., and Dean, G., 1981b, La sclerosi multipla in Sicilia: altri dati epidemiologici, in: "Atti XXII Congresso Nazionale della Società Italiana di Neurologia, Scalea, November 5-7, 1981", T. Caraceni, ed., Crippa & Berger s.p.a., Milan.

Schumacher, G. A., Beebe, G., and Kibler, R. F., 1965, Problems of experimental trials of therapy in multiple sclerosis: report by panel on the evaluation of experimental trials of therapy in multiple sclerosis, Ann.N.Y.Acad.Sci., 122:552.

RECENT DESCRIPTIVE SURVEYS ON MULTIPLE SCLEROSIS IN SARDINA AND IN THE PROVINCE OF FERRARA, ITALY

Enrico Granieri and *Giulio Rosati

Neurological Clinic of the University of Ferrara
44100, Ferrara, Italy, *Neurological Clinic of the
University of Sassari, 07100, Sassari, Italy

The Italian findings from small populations, which seem to deny the zonal concept for multiple sclerosis (MS) distribution in Europe and the more widely held interpretation of the gradient of latitude, although appealing, must certainly be reviewed with caution (Kurtzke, 1983). In fact, since MS is a disease with particular characteristics, studies based only on prevalence rather than incidence are likely to be biased by several geographic variables such as natural evolution of the disease, migratory flux, of assistance level, and accessibility to neurological centers. These problems make it difficult to compare prevalence rates for different years and areas. Moreover, owing to the size of the populations investigated, the high prevalence rates in the intensive surveys show very wide confidence intervals, which could reduce the precision rate in measuring the true frequency of MS in Italy. In particular, the large confidence intervals, due to the very few cases do not allow us to reject the idea that the true MS rate is below 30 per 100,000.

For these reasons and in the light of methodological considerations which had emerged in the previous investigations in Sardinia and in Emilia Romagna (Rosati et al., 1977; Rosati et al., 1980; Granieri and Rosati, 1982), the research groups of the Neurological Clinics of the Universities of Ferrara and Sassari, planned a further descriptive study in Barbagia, Sardinia to establish the MS incidence in the last 20 years and to estimate a new prevalence.

Barbagia covers a small area (1242 km^2, nearly 50,000 inh.) in the east-central part of Sardinia. The living population is a self contained community that was isolated for centuries and excluded from any contact with the other ethnically distinct populations occupying the island originating from southern Europe and Afro-Asiatic Mediterranean littorals. Even from a genetic point of view, Barbagia's population is different in that some of the more frequent HLA antigens found in white people are almost absent, but there are high frequencies of HLA antigens similar to those observed in Negroids and in Arabs, Turks and Lebenese (Piazza et al., 1973; Bodmer, 1973). In many ways, Barbagia appears to be appropriate for epidemiological investigations, and our groups, having used this area for years, have previously verified the suitability of this region for epidemiologic purposes.

Based on 31 MS incident cases selected according to the diagnostic criteria of Allison and Millar (1954), the mean incidence per year for the years 1961-1980 was 2.9 per 100,000 (95% confidence intervals: 1.9 - 4.1). The rate, adjusted for both age and sex versus the Italian population, was 3.2. No significant temporal trend was noted during the period considered. On October 24th, 1981 the crude prevalence rate for "probable" MS was 65.3 per 100,000 (95% confidence intervals: 44 - 93). The rate, standardized to the Italian population, was 77.9. Although incidence and prevalence rates were higher among females, no significant differences were noted between the sexes.

So, these findings, the highest figures that have yet been estimated for any community in Italy, confirm that in Barbagia MS occurs more frequently than expected in a Mediterranean area. Moreover, the high confidence intervals on the rates, both lower and upper, and the high incidence rate do in fact demonstrate that MS frequency is high. The rates give further support to the idea that MS frequency in Italy is similar to that established for most of central and northern Europe.

On the other hand, a second stage of our descriptive studies on MS in Italy includes an incidence investigation in the province of Ferrara (2632 km^2, nearly 380,000 inh.), in Emilia Romagna, northern Italy. The choice of such a large population might seem to be in contrast with what has emerged from previous Italian surveys, especially if we consider some problems which make it too difficult to obtain accurate estimates of MS frequency in our country when using large populations. However, it is necessary to bear in mind that studies on small areas can be exposed to large casual fluctuations in rates, and therefore do not permit definitive conclusions. With regard to Ferrara, a series of favorable circumstances (good level of medical organization, subdivision of the area in six homogeneous health districts, full cooperation from the doctors and the social workers engaged in the territorial services) provide us with a great amount of workable data for planning a descriptive study, in particular an incidence study, in a big population. During the years 1965-1979, 121 subjects residing in the province of Ferrara presented the onset of the disease which later developed into "probable" MS. Thus, the average annual incidence rate was 2.08 per 100,000 (2.1 if age-

and sex-standardized to the Italian population). The 95% confidence intervals are 1.7 - 2.5. During the period considered, the incidence remained stable. On October 24th, 1981 the prevalence per 100,000 for the whole province was 43.5 (95% confidence intervals: 37 - 50). The standardized rate for Italy is 40.9 per 100,000.

So, these results carried out on a large area with the same methodology used for small communities, further support the view that the frequency of MS in Italy is higher than that indicated in the literature and, on the contrary, they strengthen the idea that the risk for the disease in the Mediterranean region is similar to that found in most countries in central and northern Europe. It is possible also that in Italy in recent years there has been a real increase in the incidence of the disease, perhaps due to a progressive westernization following the Second World War, which has made the Italian life-style more similar to that in western Europe and U.S.A.

According to this latter view, a "pilot" study on the incidence of MS in the Sardinian commune of Macomer during the period 1912-1981 (Rosati et al., 1983) seems to indicate that the frequency of MS in Sardinia increased dramatically in the years following the last war. The commune of Macomer covers a small area (123 km^2, nearly 11,000 inh.) in the western part of the province of Nuoro, Sardinia situated between 39°54' and 40°13' N and 8°58' and 9°39' E. The living population is a self-contained community that remained isolated until the late forties and early fifties. In those years there was an immigration of workers and technicians from norther Italy and the Netherlands since a number of industrial activities (brewery, wool industry, cheese factory, etc.) were initiated.

The incident cases were drawn from all possible sources: archives of discharge diagnosis for the years 1912-1983 from the Sassari and Cagliari University Hospitals and from the hospitals of Nuoro and Ozieri; death certificates for the years 1911-1980 from the commune of Macomer, local section of the Italian Association for MS; and directly from the old municipal doctor of Macomer who practised in the area from 1933 to 1978. During the years 1912 through 1981, 13 (9 women and 4 men) native-born subjects living in the study area presented onset of disease which later developed into "definite" MS, according to the criteria of the Schumacher Committee (1965). All these cases had clinical onset of MS between 1952 and 1981; in contrast the disease was completely absent before 1952. During the period 1952 - 1981 the mean annual incidence was 6.3 per 100,000 (95% confidence intervals: 3.4 - 10.8). By estimating the incidence rates for 5-year intervals, the onset of MS in Macomer showed the following temporal distribution: 2.8 cases per 100,000 population in the years 1952 - 1956; 10.2 in the years 1957 - 1961; 7 in the years 1962 - 1966; 4.2 in the years 1967 - 1971; 3.9 in the years 1972 - 1976; and 1.8 in the years 1977 - 1981 (Figure 1). In 1961 the point prevalence rate was 61.5 per 100,000 population (95% confidence intervals:

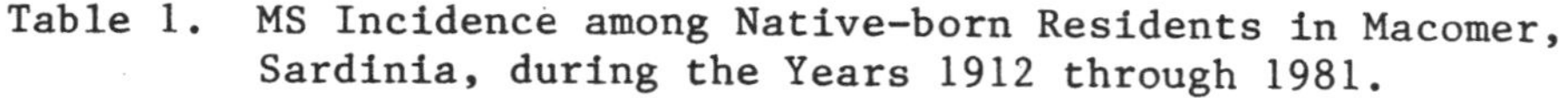

Table 1. MS Incidence among Native-born Residents in Macomer, Sardinia, during the Years 1912 through 1981.

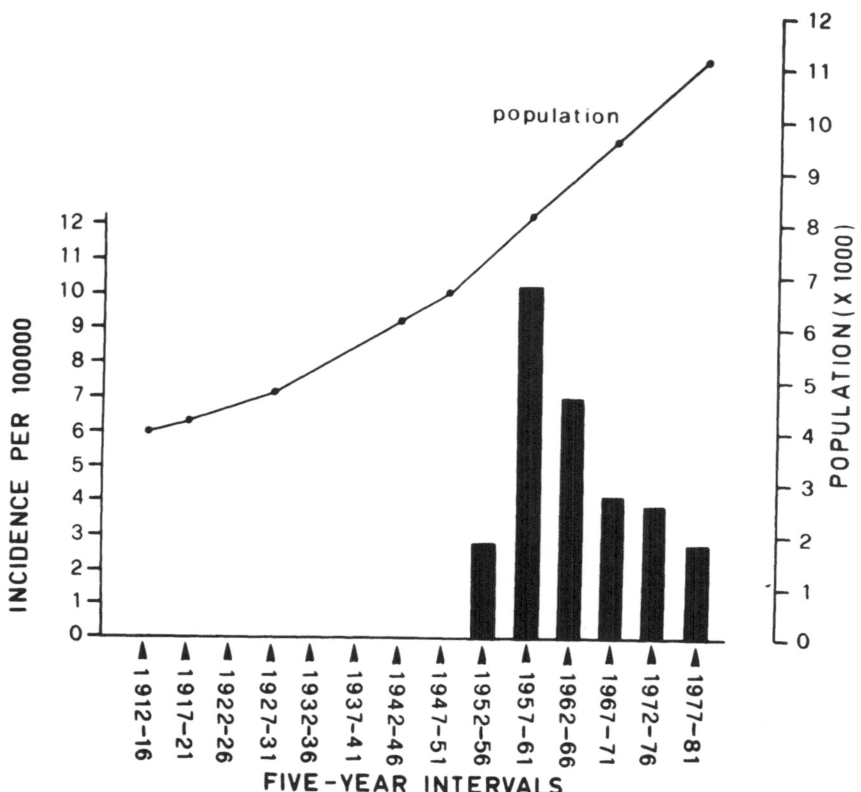

20 - 143); in 1971, 73.4 (95% confidence intervals: 30 - 151); and in 1981, 72 (95% confidence intervals: 31 - 141). In other words, MS appeared in Macomer in the early fifties, reached its highest incidence in the years 1957 to 1967, and then decreased slowly until 1981.

These findings lead to the conclusion that MS was introduced in Macomer, and probably in other parts of central Sardinia, in the years following the Second World War, when the rise of industrialization put an end to the century-old isolation of the native populations. These data, moreover, seem to confirm the hypothesis posed in the famous study by Kurtzke and Hyllested (1979) on the Faroe Islands that MS is an exogenous disease with epidemic trend.

Acknowledgement

This work is supported by the funds of the Italian Ministry of Education and partly by a grant from the Bank of Sardinia.

Our thanks to Dr Francesco Pisanu for his continued support in the planning of the study and in the gathering of the data.

REFERENCES

Allison, R. S., and Millar, J. H. D., 1954, Prevalence and familial incidence of disseminated sclerosis: a report to the Northern Ireland Hospitals authority on the results of a three year survey, Ulster Med.J., 23:1.

Bodmer, W. F., 1973, Population genetics of the HLA system: retrospect and prospect, in: "Histocompatibility testing 1972", J. Dauset and J. Colombani, eds., Munksgaard, Copenhagen.

Granieri, E., and Rosati, G., 1982, Italy: A medium- or high-risk area for multiple sclerosis? An epidemiologic study in Barbagia, Sardinia, southern Italy, Neurology, 32:466.

Kurtzke, J. F., and Hyllested, K., 1979, Multiple sclerosis in the Faroe Islands: I. Clinical and epidemiological features, Ann.Neurol., 5:6.

Kurtzke, J. F., 1983, Some epidemiological trends in multiple sclerosis, Trend Neuro.Sci., 6:75.

Piazza, A., Belvedere, M. C., and Bernoco, D., 1973, HLA variation in four Sardinian villages under differential selective pressure by malaria, in: "Histocompatibility Testing 1972", J. Dausset and J. Colombani, eds., Munksgaard, Copenhagen.

Rosati, G., Pinna, L., Granierei, E., Aiello, I., De Bastiani, P., Tola, R., Agnetti, V., and Pirisi, A., 1977, The distribution of multiple sclerosis in Sardinia, Riv.Pat.Nerv.Ment., 98:46.

Rosati, G., Granieri, E., Carreras, M., and Tola, R., 1980, Multiple sclerosis in southern Europe: A prevalence study in the socio-sanitary district of Copparo, northern Italy, Acta Neurol. Scand., 62:244.

Rosati, G., Aiello, I., Pirastru, M. I., Becciu, S., Zoccheddu, A., De Montis, G., and Mannu, L., Studio sull'andamento temporale della sclerosi multipla a Macomer dal 1912 al 1981: conferma epidemiologica della natura esogena della malattia, in: "Atti XXIII Congresso Nazionale della Societa` Italiana di Neurologia, Parma, October 20-22, 1983", in press.

Schumacher, G. A., Beebe, G., and Kibler, R. F., 1965, Problems of experimental trials of therapy in multiple sclerosis: report by panel on the evaluation of experimental trials of therapy in multiple sclerosis, Ann.N.Y.Acad.Sci., 122:552.

EVOKED POTENTIALS IN THE INVESTIGATION OF MULTIPLE SCLEROSIS

W. B. Matthews

Professor of Clinical Neurology
University of Oxford
Radcliffe Infirmary, Oxford, UK

The use of evoked potential (EP) techniques in the investigation of a number of aspects of multiple sclerosis (MS) has been a major growth industry. The technique has been used in the diagnosis of MS, in attempts at monitoring the progress or activity of the disease, including the effects of treatment, and in exploring its pathophysiology. The time is ripe for a critical appraisal of how far claims for the value of EP in the investigation of MS are justified.

DIAGNOSIS

In the evaluation of any diagnostic technique, it is first essential to demonstrate that the test under investigation is positive or abnormal in a high proportion of known cases of the disease. In the whole of medicine very few diagnostic tests have been shown to be 100% positive and much lower figures are valuable if they do not include an undue proportion of false positives. In MS there is the further complication that the duration of the disease is nearly always extremely prolonged and even if it were possible to await a fatal outcome, there may be no opportunity to confirm the diagnosis as the autopsy rate is notoriously low. Diagnosis must therefore be on clinical grounds. Every few years a new attempt is made to establish criteria on which a diagnosis of 'clinically definite' MS can be made. The latest criteria include abnormal results of laboratory tests, among these notably the results of EP recording, but this is clearly inappropriate when the objective is to assess the diagnostic value of these same techniques.

Most authors who have studied EP in the diagnosis of MS have used McAlpine's[1] criteria. These are reasonable when applied to the clinically definite category but careful perusal at once reveals that the definition of 'probable' and 'possible' cases contains important ambiguities. However, most other criteria have been devised for a different purpose, in particular for the recognition of cases suitable for trial of therapeutic agents, when it would be clearly inadmissible to include patients in whom the diagnosis was doubtful.

The first diagnostic application of EP in MS was in the use of the averaged visual potential (VEP) evoked by stimulation of each eye separately by means of an alternating chequerboard pattern[2]. Earlier attempts using the simpler stimulus of a flash of light had not produced consistent results although this technique may still be occasionally useful[3]. The actual stimulus in the shifting pattern appears to be the edges of the squares.

Halliday et al.[2] applied this technique to patients known to have had a recent episode of optic neuritis (ON). The finding of a delayed response was of great interest but was, of course, of no diagnostic value, as demyelinating disease of the optic nerve was not in doubt. The latency was measured from stimulus, that is to say the movement of the pattern, to the peak of the most prominent component of the evoked response, the so-called P100. There are obvious sources of error. The pattern shift is not instantaneous and indeed in the television screen devices now in common use the pattern develops on the screen in sequence, usually from the upper left corner, although this is not detectable to the eye. The P100 is chosen because it is the most consistent response, although certainly not the earliest component. Using this technique Halliday et al.[2] found a notable asymmetry of latency between the responses evoked from the affected eye and that from the presumed normal eye. Even at this stage, however, it should be noted that this abnormality was not present in all affected eyes, a normal response being obtained on one occasion. This may be regarded as an illustration of the failure to find 100% positive results with any diagnostic test.

These findings were confirmed in other series. Prolonged latency was found on stimulation of 22 of 28 eyes known to have been affected by optic neuritis within the preceding three months, the response from the other six eyes being normal. Asymmetry of latency was less striking, and indeed in one patient latency was more prolonged on stimulation of the eye not known to have been involved[4]. Shahrokhi et al.[5] examined 51 patients without evidence of multiple lesions but who have a history of optic neuritis at some time in the past. A response normal in all respects was obtained from three of 59 affected eyes, latency being increased in the majority of the remainder. Such results indicate that optic neuritis in the past was often associated with prolonged latency of VEP.

The next stage was to record VEP in patients with clinically definite MS not suffering from recent ON. In the original series Halliday et al.[6] found abnormal VEP in 97% of such patients. This figure has been approached or even surpassed in a number of later studies[7-9], but somewhat lower proportions are more usual[4,10-13] but still above 75%. These findings, while an essential step in the exploration of the technique, were again of no diagnostic value as the patients were already accepted as having MS. Of great importance to the extension of the technique to the examination of suspected cases is the proportion of abnormal VEP in definite cases with no clinical evidence of involvement of the visual system; that is to say, no history of ON, no abnormal pallor of the optic discs, normal acuity and color vision, the usual routine evaluation. The visual field changes in such patients, when present at all, are subtle and difficult to record.

Not unexpectedly the percentage of abnormal VEP is considerably higher in patients with a past history of ON than in those without. Even the notoriously fallible clinical sign of pallor of the temporal half of the optic disc is associated with a higher incidence of abnormal VEP[4]. Nevertheless, even in the absence of any evidence of visual involvement, present or past, VEP are abnormal in clinically definite MS in from 57 to 94% of cases[6,7]. The significance of this is that the method appears to be capable of detecting abnormality where none was otherwise suspected - the silent lesion. The importance of this finding to the diagnosis is obvious enough. Multiple lesions may be demonstrated by means other than clinical history and examination.

The technique was therefore legitimately applied to the examination of patients in whom the clinical features did not satisfy the criteria of definite MS. In the original series[6] such abnormalities were indeed found in 100% of 'probable' cases of MS and in 92% of 'possible' cases. McAlpine's criteria for the probable category are difficult to apply at all strictly and this may account for the much lower figures for abnormal VEP in such cases reported in most series, 33%[14,15], 50%[10] and 58%[4,9]. The remarkable figure of 92% of abnormalities in patients in the possible category has never been approached. By definition very few such cases will have convincing clinical evidence of optic nerve affection, so that any EP abnormalities detected are likely to be of greater diagnostic value.

If decisions are to be made affecting the management of the patient it is clearly of the utmost importance that strictly defined standards of normality are established. The technique has shown itself to be remarkably robust in that consistent results can be obtained despite variation in such factors as the size of the squares in the stimulus pattern and the luminance of the black and white areas[16]. Nevertheless it is not safe to base conclusions on normal results published by others, and each laboratory must establish normal values. Latency measurements beyond 2.5 or 3 SD above the mean are usually taken as abnormal but clearly more weight will be attached to a figure 50% greater than the mean than a latency a few milliseconds beyond the upper limit chosen. Asymmetry of latency beyond limits observed in normal series is also often accepted as indicating pathology even if latency is not unduly prolonged. Gross asymmetry of amplitude may also be abnormal but low amplitude symmetrical responses of normal latency are not uncommon in healthy subjects. Amplitude is often much greater in women and mean latency is also slightly shorter. There are unresolved difficulties in interpreting certain wave forms of unusual shape. There is little effect of age on VEP latency within the age range of patients with MS. Visual acuity below 6/12, from whatever cause, may result in an abnormal response[17].

Taking all these factors into account, a large number of reliable surveys have now been published and all demonstrate that the less certain the clinical diagnostic category the lower the rate of abnormal VEP found. This must mean either that the investigation is relatively seldom abnormal in early cases of MS or that a high proportion of patients in the lower categories of diagnostic certainty do not have the disease. A further difficulty arises, that will be discussed later in detail. What evidence exists to justify 'transferring' patients from one diagnostic category to another, implying a greater degree of certainty, on the evidence of an abnormal VEP recording? Many authors appear to have sensed this difficulty and announce the intention to follow up such patients to determine the outcome, but very few have succeeded in doing so. Others have accepted without question that such results indicate 'proven dissemination of the demyelinating process'[13]. It is, perhaps, unwise to speak of 'proof' of a biological phenomenon from indirect evidence.

The collection of data on percentage of abnormalities and the theoretical 'transferring' of blocks of patients as a result is, perhaps, rather far removed from the practical problems of the clinic. The possible diagnostic value of VEP may be considered in two common clinical problems. The first concerns the young adult who presents with a single monosymptomatic episode compatible with the onset of MS - diplopia or sensory symptoms and signs in the limbs. The immediate diagnosis of MS in such circumstances carried no immediate benefit for the patient except that other causes of, for example, a VIth cranial nerve palsy, need not be sought. A single episode of this kind does not qualify a patient for any of McAlpine's diagnostic categories but would be regarded as 'suspect' by McDonald and Halliday[18]. The frequency with which recording VEP provides evidence of an unsuspected lesion of the visual system in such patients has seldom been examined but is certainly small: three out of 39[4] and two out of 22[19] in two series.

The other clinical problem to which VEP recording might be expected to contribute is that of the patient with progressive spastic weakness of the legs and no clinical evidence of multiple lesions. It is well known that a high proportion of such patients do indeed have MS but there are unpleasant possibilities of diagnostic error, including overlooking a benign cause of spinal cord compression. Abnormal VEP in such a patient should indicate multiple lesions and therefore obviate the necessity for myelography, Blumhardt et al.[20] attempted to find the answer in 64 patients with chronic progressive paraparesis, out of a total of 131 cases of apparently isolated spinal cord disease. In the chronic progressive category VEP were abnormal in 36%, the percentage being much higher in those with symptoms for more than three years. Unfortunately for the purely scientific value of the study the clinicians involved were undoubtedly influenced by the results of VEP recording and myelography was not done in a high proportion of those in whom VEP were abnormal. This was the precise result the investigation was designed to test and it is not known how often the decision not to proceed to invasive radiology was justified.

The question of the specificity of abnormal VEP is of obvious importance. Prolonged latency may be found in hereditary ataxias[21], hereditary spastic paraplegia[21], vitamin B12 deficiency[22], the Chiari malformation[23], neurosyphilis[24] and sarcoidosis[25], all conditions that may be confused with MS. Abnormal VEP in patients with spinal cord disease and myelographically demonstrated narrowing of the cervical spinal canal may be explained by the well recognized coexistence of MS and spondylotic myelopathy[26]. Conditions that do not cause spinal cord disease but which may result in prolonged VEP latency and cause diagnostic confusion include glaucoma without visual loss[27] and diabetes without retinopathy[28]. Abnormal VEP are certainly of diagnostic use in MS if used judiciously but cannot be regarded as 'a valuable test in the early diagnosis' as was at first hoped[6].

SOMATOSENSORY EVOKED POTENTIALS (SEP)

Namerow[29] reported that the latency of the scalp recorded SEP evoked by stimulation of the median nerve was prolonged in patients with MS with sensory loss in the limb tested. In diagnostic use attention has been mainly directed to short latency potentials recorded over the neck and scalp from stimulation of main nerve trunks in the limbs[30]. By the latter method a clear potential can be recorded from the scalp but there are great difficulties in obtaining consistent records over the spinal cord above the lower dorsal region. In relaxed normal subjects and by averaging a very large number of responses an extremely small amplitude potential can sometimes be recorded from surface electrodes over the cervical spinal cord from stimulation of the tibial nerve[31]. This naturally gave rise to hopes of measurement of spinal cord sensory conduction velocity in disease but this has not proved possible. In spinal disease such potentials cannot be recorded at all but it is not possible to accept this as evidence of disease in cases of doubt as the response is often unrecordable in normal subjects.

In routine clinical practice the median nerve is stimulated and the averaged response recorded over the neck, referred to the vertex, and the short latency scalp component is also recorded. The complex series of potentials in the neck recording include an initial wave arising from the brachial plexus and if latency of this component is normal this provides good evidence of normal peripheral conduction. Naturally latency will vary according to length of arm. The site of origin of the main component of the normal cervical response is probably the dorsal column nuclei rather than the cervical spinal cord.

In MS both the cervical and early scalp potentials are often abnormal. This is not surprising as the posterior columns in the cervical spinal cord are sites of predilection for MS plaques. Small et al.[32] reported abnormal cervical SEP in 69% of clinically definite cases and Mastaglia et al.[15] in 15 of 17 such cases. The abnormality observed was usually absence of the cervical potential rather than prolonged latency. It is tempting to regard wave forms of unfamiliar shape as indicating pathology but this is almost certainly unjustified. The early scalp SEP, N20, is seldom abnormal if the cervical response is normal. Prolonged latency of this component may be observed.

In patients in whom the diagnosis of MS is uncertain, SEP are also frequently abnormal. Small et al.[32] reported abnormal SEP in 42% of probable and 52% of possible cases, the latter category including, by definition, cases of progressive spastic paraparesis. In many of these patients the abnormal SEP could not be accepted as revealing leasions in regions of the CNS not known to be involved as spinal cord disease was clinically evident. However, in this series, 21% of patients with no sensory loss in the upper limbs and no evidence of spinal cord disease had abnormal SEP. These stricter criteria are obviously sound and were adopted by Green et al.[33] and by Purves et al.[34].

The use of SEP in the early diagnosis of MS has encountered difficulties similar to those with VEP. In patients with a single symptomatic episode SEP are seldom abnormal. It is probable that many of these patients do not have MS and will never develop the disease, but there is one group in whom the probability of developing MS is high: patients with isolated optic neuritis. In a series of 40 such cases SEP from the median nerve were abnormal in only two, a far smaller proportion that the expected incidence of MS in these patients. The conclusion must be either that plaques in the cervical spinal cord are seldom present at this stage of the disease or that they are not detectable by this method.

AUDITORY EVOKED POTENTIALS

Brainstem auditory EP (BAEP) at first appeared to provide much greater precision of interpretation as each of the seven components usually recordable could be duplicated in animal experiments and attributed to activity in a closely localized site in the auditory pathway. This is in contrast to VEP and SEP where there is still much controversy concerning the origin of the potentials. Experience has shown that localization of lesions by means of BAEP is less precise than was hoped. This is scarcely surprising as pathological processes are seldom as sharply defined as those produced in the laboratory and spreading effects from pressure, oedema and other factors commonly occur. A further disadvantage of BAEP is their remarkable sensitivity to variations in stimulus, method of recording and status of the subject examined[36].

In the diagnosis of MS it has been found that BAEP are less often abnormal in clinically definite MS than VEP or SEP, although figures have varied widely between different series. Chiappa et al.[37] found only 57% of abnormalities in definite cases with clinical evidence of brainstem disease, whereas Stockard et al.[38] reported 77% of abnormalities in probable cases who had never had brainstem symptoms. In a further report[36] they described results in a large series of patients with a single lesion suggestive of MS, either ON or myelopathy. When examined in the first attach only 10% of patients had abnormal BAEP. Most exceptionally an adequate follow up study was mounted and it was found that all those with abnormal BAEP in the first attack developed MS within three years.

MULTIPLE EP RECORDINGS

The availability of several different methods of recording EP naturally led to the application of such multiple examinations in the hope of increasing the proportion of abnormalities detected. The combined use of VEP, SEP by various techniques, and BAEP has been described by a number of authors[19,33,34,39,40,41,42,43]. There is no doubt that if all three modalities are used abnormalities may be found in individual patients by one method alone. Many authors have claimed that SEP is the most useful in this respect but the criteria employed have been defective. Abnormal SEP cannot be claimed as demonstrating a silent lesion in a patient with spastic weakness of both legs, even if no sensory loss is found, as disease of the spinal cord is already evident.

It may be doubted whether the submission of a patient with suspected MS to two hours' intensive and sometimes uncomfortable and often frightening investigation in an EP laboratory - the so-called routine MS workup - can be justified by results. Papers claiming success have mainly been written by clinical neurophysiologists who are not in charge of the patients.

CAN ABNORMAL EP INDICATE SILENT LESIONS IN MS?

The diagnostic use of EP has been largely built on assumptions; first that abnormalities can result from asymptomatic plaques, and secondly that the discovery of such abnormalities is diagnostic of MS. Anatomical verification of the first assumption is almost entirely lacking. I recorded SEP from a woman with very mild MS[44]. She had experienced sensory symptoms in the right limbs in the past but at the time I saw her there was no sensory loss, although the right leg was weak and spastic. Shortly afterwards she died suddenly from a subarachnoid hemorrhage. At autopsy a plaque was found encroaching on the dorsal root entry zone and on the posterior columns in the cervical spinal cord. This could confidently be accepted as the cause of the abnormal SEP in the absence of sensory loss.

That the second assumption is correct requires prolonged follow up of patients in whom EP abnormalities have appeared to indicate the presence of silent lesions allowing their 'transfer' to some more definite diagnostic category. Do such patients actually have MS? Follow up of such patients is extremely difficult. They are young adults who have mostly had transient symptoms and who may not know that MS has even been suspected. They are not inclined to waste time attending hospital. Few such studies have been reported. Deltenre et al.[45] examined 133 patients suspected of having MS by all three techniques and followed them up for up to four years. In 27 some diagnosis other than MS was established and in five of these abnormal EP indicating a silent lesion had been found - false positive for MS. Disturbingly, BAEP were abnormal in two patients with spondylotic myelopathy. Of the 44 patients who developed clinically definite MS, 36 had abnormal EP when first examined. Of the 62 in whom no diagnosis had been reached on follow up 12 had abnormal EP of uncertain significance.

In a personal study[19] containing a higher proportion of patients with a single episode of neurological symptoms, 17 of 56 patients developed definite MS within three years and of these nine had abnormal EP when first seen. Absence of such abnormalities was of no prognostic significance. The lack of follow up studies should be remedied. For example, abnormal BAEP have been reported in patients with ON[46] but it is not known whether MS developed.

PRACTICAL DIAGNOSTIC VALUE

A serious contribution to the management of patients would be methods of diagnosis that would obviate the use of invasive radiology, particularly myelography. Mastaglia et al.[47] specifically examined this aspect. Of 129 patients with probable or suspected MS such procedures were avoided in 23 after taking into account neurophysiological findings and perhaps should have been avoided in a further 24. No follow up was reported so it is not known whether the omission of radiology was always justified by the event.

The most useful application of EP in MS is still that of recording VEP in patients with disease of the spinal cord. Prolonged latency is strong evidence of more widespread disease, and the probability of MS is much increased if the abnormalities are markedly asymmetrical.

MONITORING

There is a serious need for an objective method of monitoring the activity of the disease process in MS and of measuring the change in clinical status. This is particularly relevant to the assessment of methods of treatment. The concept that these objectives might be attained by serial recordings of EP is at first sight attractive but contains a fundamental flaw. The value of EP in diagnosis lies in their ability to demonstrate persistent abnormality in the absence of relevant clinical signs, and in apparently quiescent disease. This property is almost the reverse of those required for an efficient monitor. What is needed is a measurement that will reliably indicate relapse and remission. The well known lack of relationship between lesions and symptoms in MS naturally renders the interpretation of any such indications extremely difficult. To demonstrate the persistence of a plaque or even an increase in its neurophysiological effect, in the absence of relevant symptoms, cannot readily be translated into persistent 'activity' of disease or deterioration of the clinical state. Nevertheless it was essential to explore these possibilities.

Detailed serial recordings of VEP and SEP were performed in a small number of patients by Matthews and Small[48]. Records were obtained at monthly intervals and correlated with the clinical state. In one particular monitoring appeared to be successful as sudden and persistent increase in latency of VEP accompanied brief slight changes in visual acuity. SEP, in contract, either remained constant if recordings were normal or, if abnormal, fluctuated widely without relationship to any change in clinical features. Similar fluctuations in abnormal BAEP in MS were observed by Robinson and Rudge[46].

Less detailed serial records repeated once or twice after a prolonged period in patients with definite MS have shown that with the passage of time the mean latency of a number of components increases in relation to the total disability rather than to clinical abnormalities in the systems tested[49]. In contrast, Confavreux et al.[50] found that the latency of VEP, measured at an interval of a year, was not related to the degree of disability, but increased in relapse, whether the visual system was clinically involved or not. Visual function does not, however, seem to have been examined in any detail.

Likosky and Elmore[51], among others, have claimed that fluctuations in EP values in the absence of any clinical change indicate subclinical relapses. This may be so, but is impossible to confirm without correlation with some other indication of disease activity such as myelin basic protein in the CSF or new plaques detected on NMR scan. Detailed monitoring of individual patients for disease activity and clinical status does not seem practicable, partly because we do not know the significance of the changes observed.

PATHOPHYSIOLOGY

Halliday and McDonald[52] interpreted the prolonged latency of VEP in ON as due to persistent demyelination and the reduction in amplitude to conduction block resulting from reversible swelling and oedema. The amplitude was shown to return rapidly to normal with restoration of visual acuity, while latency remained prolonged indefinitely. The role of oedema in the production of symptoms in MS is somewhat speculative and certainly amplitude of VEP can remain depressed long after an acute attack of ON.

It is well known that an increase in environmental temperature will cause reduction of visual acuity in MS patients with optic nerve lesions. This is in conformity with the observed effect of heat in causing conduction block in demyelinated fibers at physiological temperatures. In such patients raising body temperature results in a reduction in amplitude of VEP and cervical SEP with usually no effect on latency[53]. It therefore appears probable that persistent reduction in amplitude is due to conduction block of a proportion of fibers due to demyelination in contrast to any rapid changes noted in acute ON.

It is customary to speak of delay of the P100 in the visual system or of wave V in the BAEP but it is of course by no means certain that these waves are indeed the 'same' waves as those present in the normal, simply occurring later because of slowed conduction in the white matter. This may indeed to be but it has never been easy to explain, for example, a 40 msec delay in latency of the P100 in terms of slowed conduction through the optic nerve. McDonald[54], however, calculated that a 1 cm long plaque could introduce a 25 msec delay. The nature of the recorded response must surely be susceptible to such factors as conduction block in faster fibers and dispersion of the potential due to unequal slowing in different fibers.

The mechanism of remission in MS remains mysterious but is clearly central to any hope of restoration of function by therapeutic means. There is really no certain evidence of remyelination in MS and this is in accord with the persistence of abnormal EP, even when function has somehow returned to normal. Restitution of normal latency of VEP has occasionally been observed[48] but usually in relation to significant improvement in acuity following acute ON. We[55] were recently able to demonstrate marked prolongation of latency, with normal vision, persisting for 3½ years at least, but returning to normal five years after the episode of ON. This has proved to be a rare, but not isolated, occurrence on similar extended follow up, but if prolonged latency is due to demyelination, this finding suggests that delayed remyelination may occur.

CONCLUSION

The commercial availability of excellent equipment for recording EP has naturally led to excessive use of these techniques in the diagnosis of MS. Early cases can seldom be detected and in more advanced disease such diagnostic aids are less often needed. There are many technical and clinical pitfalls for the inexperienced. In contrast, the potentialities of EP for investigating the pathophysiology of MS have not been fully exploited.

REFERENCES

1. D. McAlpine, "Multiple Sclerosis: A Re-Appraisal," D. McAlpine, C.E. Lumsden and E.D. Acheson, eds., Churchill Livingstone, Edinburgh (1972).
2. A. M. Halliday, W. I. McDonald and J. Mushin, Delayed visual evoked response in optic neuritis, Lancet, 1:982 (1972).
3. A. Neetens, Y. Hendrata and J. van Rompaey, Pattern and flash visual evoked responses in multiple sclerosis, J.Neurol., 220:113 (1979).
4. W. B. Matthews, D. G. Small, M. Small and E. Pountney, Pattern reversal evoked visual potential in the diagnosis of multiple sclerosis, J.Neurol.Neurosurg.Psychiat., 40:1009 (1977).
5. F. Shahrokji, K. H. Chiappa and R. R. Young, Pattern shift visual evoked responses: two hundred patients with optic neuritis and/or multiple sclerosis, Arch.Neurol., 35:65 (1979).
6. A. M. Halliday, W. I. McDonald and J. Mushin, Visual evoked response in diagnosis of multiple sclerosis, Brit.Med.J., 4:661 (1973).

7. M. Hennerici, D. Wenzel and H. J. Freund, The comparison of small-size rectangle and checkerboard stimulation for the evaluation of delayed visual evoked responses in patients suspected of multiple sclerosis, Brain, 100:119 (1977).
8. F. Chain, J. Mallecourt, M. Leblanc and F. Lhermitte, Apport de l'enregistrement des potentials évoqués visuels au diagnostic de la sclérose en plaques, Rev.Neurol., 133:81 (1977).
9. W. Trojaborg and E. Petersen, Visual and somatosensory evoked cortical potentials in multiple sclerosis, J.Neurol. Neurosurg.Psychiat., 42:323 (1979).
10. D. W. K. Collins, J. L. Black and F. L. Mastaglia, Pattern-reversal visual evoked potentials: method of analysis and results in multiple sclerosis, J.Neurol.Sci., 36:83 (1978).
11. P. Asselman, D. W. Chadwick and C. D. Marsden, Visual evoked responses in the diagnosis and management of patients suspected of multiple sclerosis, Brain, 98:261 (1975).
12. F. Shahrokji, K. H. Chiappa and R. R. Young, Pattern shift visual evoked responses, Arch.Neurol., 35:65 (1978).
13. K. Lowitzsch, Pattern-evoked visual potentials in 251 multiple sclerosis patients in relation to ophthalmological findings and diagnostic classification, in: "Progress in Multiple Sclerosis Research," H.J. Bauer, S. Poser and G. Ritter, eds., Springer, Berlin (1980).
14. G. G. Celesia and R. F. Daly, VECA - a new electrophysiological test for the diagnosis of optic nerve lesions, Neurology, 27:637 (1977).
15. F. L. Mastaglia, J. L. Black and D. W. K. Collins, Visual and spinal evoked potentials in diagnosis of multiple sclerosis, Brit.Med.J., 2:732 (1976).
16. A. M. Halliday, Event-related potentials and their diagnostic usefulness, in: Motivation, Motor and Sensory Processes of the Brain, H.H. Kornhuber and L. Deecke, eds., Elsevier, Amsterdam (1980).
17. D. W. K. Collins, W. M. Carroll, J. L. Black and M. Walsh, Effect of refractive error on the visual evoked response, Brit.Med.J., 1:231 (1979).
18. W. I. McDonald and A. M. Halliday, Diagnosis and classification of multiple sclerosis, Brit.Med.Bull., 33:4 (1977).
19. W. B. Matthews, J. R. B. Wattam-Bell and E. Pountney, Evoked potentials in the diagnosis of multiple sclerosis: a follow-up study, J.Neurol.Neurosurg.Psychiat., 45:303 (1982).
20. L. D. Blumhardt, G. Barrett and A. M. Halliday, The pattern visual evoked potential in the clinical assessment of undiagnosed spinal cord disease, Adv.Neurol., 32:463 (1982).
21. L. Pedersen and W. Trojaborg, Visual auditory and somatosensory pathway involvement in hereditary verebellar ataxia, Friedreich's ataxia and familial spastic paraplegia, Electro enchep.Clin.Neurophysiol., 52:283 (1981).
22. E. J. Fine and M. Hallett, Neurophysiological study of subacute combined degeneration, J.Neurol.Sci., 45:331 (1980).

23. A. M. Halliday and W. I. McDonald, Pathophysiology of demyelinating disease, Brit.Med.Bull., 33:21 (1977).
24. B. Conrad, R. Benecke, H. Müsers, H. Prange and W. Behrens-Baumann, Visual evoked potentials in neurosyphilis, J.Neurol. Neurosurg.Psychiat., 46:23 (1983).
25. L. J. Streletz, R. A. Chambers, H. B. Sung and H. L. Israel, Visual evoked potentials in satcoidosis, Neurology, 31:1545 (1981).
26. W. R. Brain and M. Wilkinson, The association of cervical spondylosis and disseminated sclerosis, Brain, 80:456 (1957).
27. C. Huber and T. Wagner, Electrophysiologic evidence for glaucomatous lesions in the optic nerve, Ophthalmol.Res., 10:22 (1978).
28. K. Puvanendran, G. Devathasan and G. Wong, Visual evoked responses in diabetes, J.Neurol.Neurosurg.Psychiat., 46:643 (1983).
29. N. S. Namerow, Somatosensory evoked responses in multiple sclerosis patients with varying sensory loss, Neurology, 18:1197 (1968).
30. W. B. Matthews, M. Beauchamp and D. G. Small, Cervical somatosensory responses in man, Nature, 252:230 (1974).
31. S. J. Jones and D. G. Small, Spinal and sub-cortical evoked potentials following stimulation of the posterior tibial nerve in man, Electroenceph.Clin.Neurophysiol., 44:299 (1978).
32. D. G. Small, W. B. Matthews and M. Small, The cervical somatosensory evoked potentials in the diagnosis of multiple sclerosis, J.Neurol.Sci., 36:147 (1978).
33. J. B. Green, R. Price and S. G. Woodbury, Short-latency somatosensory evoked potentials in multiple sclerosis: comparison with auditory and visual evoked potentials, Arch.Neurol., 37:630 (1980).
34. S. J. Purves, M. D. Low, J. Galloway and B. Reeves, A comparison of visual, brainstem auditory and somatosensory evoked potentials in multiple sclerosis, Acta Neurol.Scand., 62:220 (1981).
35. W. B. Matthews, Somatosensory evoked potentials in retrobulbar neuritis, Lancet, 1:443 (1978).
36. J. J. Stockard, E. J. Stockard and F. W. Sharborough, Brainstem auditory evoked potentials in neurology: methodology, interpretation, clinical application, in: "Electrodiagnosis in Clinical Neurology," M.J. Aminoff, ed., Churchill Livingstone, New York (1980).
37. K. H. Chiappa, J. L. Harrison, E. B. Brooks and R. R. Young, Brainstem auditory evoked responses in 200 patients with multiple sclerosis, Ann.Neurol., 7:135 (1980).
38. J. J. Stockard, E. J. Stockard and F. W. Sharborough, Detection and localisation of occult lesions with brainstem auditory responses, Mayo Clin.Proc., 52:761 (1977).
39. P. Deltenre, A. Vercruysse, C. van Nechel, P. Ketelaer, A. Capon, F. Colin and J. Manil, Early diagnosis of multiple

sclerosis by combined multimodal evoked potentials, J.Biomed. Engineering, 1:17 (1979).
40. K. H. Chiappa, Pattern shift visual brainstem auditory, and short latency somatosensory evoked potentials in multiple sclerosis, Neurology, 30, 7 pt 2:110 (1980).
41. M. Kjaer, The value of brainstem auditory, visual and somatosensory evoked potentials and blink reflexes in the diagnosis of multiple sclerosis, Acta Neurol.Scand., 62:220 (1980).
42. S. Khoshbin and M. Hallett, Multiple evoked potentials and blink reflex in multiple sclerosis, Neurology, 31:138 (1981).
43. W. Tackmann, H. Strenge, R. Barth and A. Sojka-Raytscheff, Evaluation of various brain structures in multiple sclerosis with multimodality evoked potentials, blink reflex and nystagmography, J.Neurol., 224:33 (1980).
44. W. B. Matthews and M. Esiri, Multiple sclerosis plaque related to abnormal somatosensory evoked potentials, J.Neurol. Neurosurg.Psychiat., 42:940 (1979).
45. P. Deltenre, C. van Nechel, A. Vercruysse, S. Strul, A. Capon and P. Kietelaer, Results of a prospective study on the value of combined visual, somatosensory, brainstem auditory evoked potentials and blink reflex measurements for disclosing subclinical lesions in suspected multiple sclerosis, Adv.Neurol., 32:463 (1982).
46. K. Robinson and P. Rudge, The use of the auditory evoked potential in the diagnosis of multiple sclerosis, J.Neurol.Sci., 45:235 (1980).
47. F. L. Mastaglia, J. L. Black, L. A. Cala and D. W. K. Collins, Electrophysiology and avoidance of invasive neuroradiology in multiple sclerosis, Lancet, 1:144 (1980).
48. W. B. Matthews and D. G. Small, Serial recording of visual and somatosensory evoked potentials in multiple sclerosis, J.Neurol.Sci., 40:11 (1979).
49. J. C. Walsh, R. Garrick, J. Cameron and J. G. McLeod, Evoked potential changes in clinically definite multiple sclerosis, J.Neurol.Neurosurg.Psychiat., 45:494 (1982).
50. Ch. Confavreux, F. Mauguière, J. Courjon, G. Aimard and M. Devic, Evolution des potentials évoqués visual dans la sclérose en plaques, Rev.Neurol., 137:121 (1981).
51. W. Likosky and R. S. Elmore, Exacerbation detection in multiple sclerosis by clinical and evoked potential techniques: a preliminary report, Adv.Neurol., 32:535 (1982).
52. A. M. Halliday and W. I. McDonald, Pathophysiology of multiple sclerosis, Brit.Med.Bull., 33:21 (1977).
53. W. B. Matthews, D. J. Read and E. Pountney, Effect of raising body temperature on visual and somatosensory evoked potentials in patients with multiple sclerosis, J.Neurol. Neurosurg.Psychiat., 42:250 (1979).
54. W. I. McDonald, Pathophysiology of conduction in central nerve fibres, in: "Visual Evoked Potentials in Man," J.E. Desmedt, ed., Clarendon Press, Oxford (1977).

55. W. B. Matthews and M. Small, Prolonged follow up of abnormal visual evoked potentials in multiple sclerosis: evidence for delayed recovery, J.Neurol.Neurosurg.Psychiat., 46:639 (1983).

DISCUSSION

Ellsworth C. Alvord, Jr.
Seattle, Washington

I am delighted to learn from Professor Matthews that evoked potentials have more variability in MS than my reading of the literature had led me to believe. As I mentioned in my address on the opening day, we have been interested in using the evoked potential technique in EAE partly to confirm objectively our relatively crude neurological examinations of monkeys, partly to detect subclinical changes that might better define the onset of EAE and partly to follow the course during remissions and exacerbations to see if EAE and MS resemble each other or not.

The definition of the onset of a disease obviously depends on the techniques used to evaluate the disease: clinical, histological, chemical, physiological, etc. In our experiments evaluating various treatments of experimental allergic encephalomyelitis (EAE), we have generally used purely clinical observations to define the onset of mild, moderate or severe EAE (Shaw et al., 1976; Alvord et al., 1979b). In attempts to make more objective observations not only to corroborate our clinical observations but also to be compared with similar observations in humans with multiple sclerosis (MS) and other diseases, we have been studying changes in the cerebrospinal fluid (CSF) and in evoked potentials (Alvord et al., 1979a; Hruby and Alvord, 1980; Slimp and Alvord, 1982). We have used measurements of evoked cortical potentials to visual and somatosensory stimulation and evoked brainstem potentials to auditory stimulation as objective in vivo indicators of the probable sites of certain lesions; measurements of albumin and immunoglobulin (IgG) in the CSF as indicators of the state of the blood-CSF barrier; and measurements of myelin basic protein (BP) and anti-BP antibodies in the CSF as indicators of damage to myelin and immunologic response to this antigenic stimulation.

Male monkeys (macaca fascicularis), weighing 2-3 Kg, were injected in the hind foot pads with 0.1 ml of a water-in-oil emulsion containing 5 mg BP (pig or monkey) and 0.5 mg heat-killed mycobacterium tuberculosis. Each animal was obtained from the University of Washington Regional Primate Center, where the animals were also quarantined and housed. The Center conforms to the National Institutes of Health Standards for the care and use of experimental animals.

All animals were observed daily, one half especially carefully as we were looking for the first suggestive signs of EAE, such as abnormal behavior, anorexia or enlarged pupils, at which time evoked cortical potentials were recorded, CSF was obtained by lumbar puncture and blood was obtained from the femoral vein for the preparation of serum. At the appearance of definite signs for EAE, such as blindness, weakness or ataxia, most of the animals were treated with BP and/or dexamethazone by implanting subcutaneously one or more osmotic minipumps (Alza Corporation, 950 Page Mill Road, Pal Alto, California 94304). The animals remained under close clinical observation and evoked potentials and CSF samples were taken repeatedly throughout the periods of recovery and of any subsequent relapse.

Analyses of evoked potentials have been described in detail elsewhere (Kraft and Slimp, 1983). For the purpose of this presentation the latencies and amplitudes obtained in normal monkeys were averaged, the standard deviations calculated, and abnormalities considered definite if they exceeded 3x SD and questionable if between 2x and 3x SD.

In the 31 monkeys we observed 50 attacks of EAE, in 27 of which the observations of CSF and evoked potentials could be simultaneously correlated with the clinical observations at the onset of the attach. In 17 attacks the recordings were actually on the same day, and in the initial tabulations these were kept separate. An additional 4 attacks had recordings within the next day and another 6 within 2 to 3 days; since these were very similar to the first 17, they were combined in the tables which follow. The day of onset was defined as that day when any definite laboratory or clinical abnormality was recorded.

Analysis of these 27 attacks (Table 1) reveals that 20 were characterized by definite abnormalities in evoked potentials and CSF. Of these, 10 animals were definitely abnormal clinically but 9 were only questionably abnormal clinically and 1 actually appeared normal. Evoked potentials were definitely abnormal at the onset of EAE in 20 (74%), CSF abnormal in 25 (93%) and clinical abnormalities definitely present in only 10 (37%). If questionable clinical abnormalities are included, this last increases to 21 (78%).

The specific abnormalities of the evoked cortical potentials are analyzed in Table 2. The expected changes (i.e. an increase in latency and/or decrease in amplitude) were most commonly observed in the auditory and leg systems, whereas the reverse (facilitated or paradoxical) changes (i.e. a decrease in latency and/or increase in amplitude) were most often seen in the visual system. The visual system was least often normal, the auditory most. All 4 of the systems tested were simultaneously abnormal 5 times, 3 systems 10 times, 2 systems 3 times and 1 system 9 times.

Table 1. Correlation of clinical, electrophysiological and CSF observations at the onset in each of 27 attacks of EAE.

Evoked potentials	CSF	Clinical Observations 0	?	+	Total
N	N				0
	Abn.		1		1
?	N				0
	Abn.	4			4
Abn.	N	1	1		2
	Abn.	1	9	10	20
Total		6	11	10	27

Table 2. Abnormalities in evoked potentials at onset of EAE.

	Evoked cortical potentials* visual	auditory	arm	leg
A. No response	2			
Increased latency	2(3)	10(2)	7(2)	10(1)
Decreased amplitude	2(2)	1(1)	2(2)	2(3)
B. Decreased latency	7(3)		1(0)	1(2)
Increased amplitude	1(0)		4(2)	1(2)
C. Normal	10	14	11	10

*The first number represents definite, the number in parentheses questionable abnormalities. Definite abnormalities exceeded 3x SD, and questionable abnormalities between 2x and 3x SD.

A rather remarkable coincidence appears from these studies, namely that about 75% of attacks of EAE were first detectable by each of 4 of the 5 major techniques employed: clinical observations (including questionable signs), evoked potentials, anti-BP antibodies in the CSF and damage to the BBB. However, if one eliminates the subtle or questionable clinical signs, only 37% of attacks were defined clinically. The 5th measure, an increase in BP in the CSF, occurred at the onset of EAE in only about 25%, but this may relate to the early appearance of anti-BP antibodies precipitating most of the BP from the CSF.

Such variability in definition of the onset of EAE can be understood most easily when one recognizes that EAE is a multifocal disease, not a diffuse one, in which different parts of the CSC-CSF can be rather independently affected. On the one hand, anyone who has looked at the histologic preparations in EAE must admit this variability; on the other hand, anyone who regards BP as occurring exclusively in myelin must admit to being surprised that the blood-CSF barrier may be involved without significant involvement of the white matter in which the tissue-specific antigen is located.

With regard to recovery from EAE, Dr. Jefferson Slimp tells me that of 23 occurrences of clinical EAE with clinical recovery thus far studied in 17 monkeys, over half of these showed concomitant improvement in evoked potentials. About one-fourth showed no change and a few actually worsened. Improvement occurred equally often in the four sensory systems tested; however, the brainstem auditory evoked potentials became worse during recovery three times as often as any other system. Six bouts of EAE were followed by complete recovery of all evoked potentials to normal. Quite often this involved three or four sensory systems. Of the cases with partial improvement, however, it was characteristic that only one system returned to normal.

Thus, while the clinical recovery from EAE may appear to be complete, evoked potentials may remain abnormal in as many as two-thirds of the animals. Such subclinical abnormalities should not be surprising. The variable nature of the improvement of the evoked potentials emphasizes again that EAE is a multifocal disease with significant similarities with MS.

REFERENCES

Alvord, E. C., Jr., Hruby, S., and Sires, L. R., 1979a, Degradation of myelin basic protein by cerebrospinal fluid: Preservation of antigenic determinants under physiological conditions, _Ann.Neurol._, 6:474-482.

Alvord, E. C., Jr., Shaw, C. M., and Hruby, S., 1979b, Myelin basic protein treatment of experimental allergic encephalomyelitis in monkeys, _Ann.Neurol._, 6:469-473.

Hruby, S., and Alvord, E. C., Jr., 1980, Studies on the proteolytic activity of the cerebro-spinal fluid in multiple sclerosis and other neurological diseases, _Clin.Chem._, 26:1013.

Kraft, G. H., and Slimp, J. C., 1984, Electrophysiological monitoring of experimental allergic neuritis and experimental allergic encephalomyelitis, _in_: "Experimental Allergic Encephalomyelitis: A Useful Model for Multiple Sclerosis," E.C. Alvord, Jr., M.W. Kies, and A.J. Suckling, eds., Alan R. Liss, Inc., NYC.

Shaw, C. M., Alvord, E. C., Jr., and Hruby, S., 1976, Treatment of experimental allergic encephalomyelitis in monkeys. I. Clinical studies, _in_: "The Aetiology and Pathogenesis of the Demyelinating Diseases," H. Shiraki, T. Yonezawa, and Y. Kuroiwa, eds., Japan Science Press, Tokyo, pp. 367-376.

Slimp, J. C., and Alvord, E. C., Jr., 1982, Somatosensory visual, and auditory evoked potentials in monkeys with experimental allergic encephalomyelitis (EAE), _Neurosci.Abst._, 8:351.

CLINICAL IMPLICATIONS OF STUDIES INVOLVING CEREBROSPINAL FLUID T CELL SUBPOPULATIONS

O. J. Kolar, P. H. Rice, D. C. Bauer, R. J. Defalque, C. F. Danielson, M. R. Farlow and J. H. Wright

Multiple Sclerosis Research Laboratory
Departments of Neurology, Microbiology and Immunology, and Anesthesiology
Indiana University School of Medicine and the Department of Computing Services, Indiana University-Purdue University at Indianapolis, USA. 46223

INTRODUCTION

Multiple sclerosis (MS) is a demyelinating central nervous system (CNS) affliction manifested by exacerbations and remissions of clinical symptoms.

It has been recognized that the progressive course of MS may be associated with changes in the proportion of peripheral blood (PB) and cerebrospinal fluid (CSF) T cell subpopulations and B cells.

In PB, an exacerbation of MS may be reflected by decreased percentages of T cells,[1,2] Fcγ cells,[3] and T-suppressor-cytotoxic (OKT_8+) cells[4]. The T helper-inducer (OKT_4+)/OKT_8+ cell ratio may be increased[4,5,6] and there may be an increased percentage of B cells[7] and/or their clonal restriction[8].

In CSF, the relapse and/or progressive course of MS is manifested by an increased percentage of T cells[9,6] a decreased percentage of Fcγ cells[10] and an increased percentage of OKT_4 cells [11,12]. The percentage of OKT_8+ cells may be decreased with an increased OKT_4+/OKT_8+ cells ratio[13]. Sandberg-Wollheim[6], in her series, found, in active MS, decreased percentage of CSF OKT_8+ cells, but the reduction was not satistically significant.

Weiner et al.,[14] on examining CSF in 21 MS patients, and Hauser et al.,[15] on CSF studies in 40 MS patients did not establish characteristic changes in CSF phenotypes which could be related to changes in circulating OKT_8+ cells or to disease activity. In their series, Hauser and his coworkers[15] did not find evidence that in patients with low numbers of PB OKT_8+ cells, an accumulation of OKT_8+ cells would occur in CSF. In four age and sex matched controls followed with six MS patients over a four to six month period at weekly intervals, Hauser et al.,[16] did not establish abnormalities in PB OKT_8+ or OKT_4+ cells. In two of the MS patients studied, an elevated PB OKT_4+/OKT_8+ cell ratio was associated only with fatigue. Huddlestone and Oldstone[17] noticed increased numbers of PB Fcγ cells and/or OKT_8+ cells in MS patients in remission following an acute episode of MS.

In this presentation, we are comparing findings in our MS patients with the above mentioned observations.

MATERIALS AND METHODS

In 132 patients with definite MS (92 females, average age 38.7 ± 9.7 years and 40 males, average age 38.6 ± 9.3 years) the percentage of PB T, Fcγ, OKT_8+, OKT_4+, and B cells was determined. In 87 of these patients, more than one examination was performed (4.15 ± 2.21 examinations per one patient) over a period of time extending over 13.5 ± 8.1 months. In 44 MS patients with progressive neurological symptomatology observed up to one month prior to the CSF examination, the percentages of CSF OKT_8+, OKT_4+, T and/or Fcγ cells were also established.

The generally used techniques applied on examination for CSF T cell subpopulations were modified to obtain the maximum number of preserved mononuclear cells from the 6-8 ml of CSF submitted for a routine CSF examination.

As soon as possible following the spinal tap, the CSF specimen, in a 15 ml plastic conical centrifuge tube, was centrifuged at 400 g for 30 minutes at 5° C. One ml of CSF was left at the bottom of the tube and the remaining CSF was thereafter carefully removed. The cells collected at the bottom of the tube were then gently re-suspended and divided into four labeled 12 x 75 mm tubes using a serologic pipet. 0.15 ml of AET-treated sheep red blood cells were added in the tube for examination of rosetting T cells. 0.15 ml of ox IgG coated red blood cells were placed in the tube for examination of the Fcγ cells. Following spinning at 200 g at 5°C for five minutes, the tubes were kept on ice for one hour. Thereafter, the supernatant was carefully removed, leaving 0.3 ml of the cell suspension at the bottom of the tube. The cells were subsequently gently resuspended and rosetting cells were counted in a hemacytometer. For examination of the T suppressor-cytotoxic cells and the helper cells, 5 microliters of reconstituted Ortho-mune OKT_8 and OKT_4

antibody solutions (Ortho Pharmaceutical Co., Raritan, NJ) were added to the remaining two tubes. Following incubation on ice for 30 minutes, one drop of a goat-anti-mouse IgG fluorescein conjugated serum (Meloy Lab., Inc., Springfield, VA) was placed in each tube using a Pasteur pipette. After incubation on ice for an additional 30 minutes, 2 ml of PBS were added to each tube and the tubes were centrifuged at 300 g for 10 minutes at 5°C. The supernatant was subsequently removed leaving approximately 20 microliters of the cell suspension in the tube. Following gentle resuspension of cells, 10 microliters of the cell suspension were released from the pipette onto a coverslip. The labeled slide was then carefully placed over the coverslip which was sealed with vasoline and the slide was read without delay.

In case a larger amount of CSF is submitted for examination, or there is a special reason to establish subgroups of CSF B cells, fluorescein conjugated antisera to immunoglobulins G, A, M and D and/or to kappa and lambda light chains may be used similarly as the above mentioned solutions of monoclonal antibodies.

The median number of CSF mononuclear cells evaluated on determination of the percentages of T and Fcγ cells was 67 and of the OKT_4+ and OKT_8+ cells was 50. The CSF specimen examination also included cell count and cytomorphology; cellulose polyacetate protein electrophoresis, immunoelectrophoresis using antisera to human serum, Fab fragments of IgG, kappa and lambda chains, and determination of the IgG concentration and the IgG/Albumin ratio as previously described[18]. Protein electrophoresis, immunoelectrophoresis and examination of IgG, IgM and IgA levels in the corresponding serum specimens were also performed.

The percentages of PB T, Fcγ, OKT_8+, OKT_4+ and B cells were determined in 30 control subjects (average age 37.5 ± 9.9 years). In 13 of them, more than one examination (2.8 ± 1.0) was performed over 15.6 ± 8.2 months. In addition, the percentages of CSF T cells (N=22, 86.59 ± 8.16), Fcγ cells (N=20, 19.70 ± 10.64), OKT_4+ (N=12, 47.33 ± 16.04), and OKT_8+ cells (N=20, 30.97 ± 13.00) were determined in a control group of individuals without neuropsychiatric symptoms and with normal CSF cell count, total proteins, CSF cytomorphology and protein electrophoresis. For statistical evaluation, Spearman and Kendall correlation coefficients, Wilcoxon, Kruskal-Wallis, Van Der Waerden, Median and Savage tests were applied.

RESULTS

In the controls without neuropsychiatric symptoms, the percentages of CSF T cells and Fcγ cells were found to be higher as compared to PB in control subjects (T cells 73.36 ± 8.33; Fcγ cells 12.56 ± 6.76).

In our series of MS patients with progressive neurological symptomatology (N=44) the average CSF cell count was 11.09 ± 9.24 white blood cells per cubic millimeter. The average percentage of CSF lymphocytes was 91.21 ± 5.92. In the MS patients studied, the percentage of CSF T cells (90.23 ± 8.14) was higher compared to the corresponding PB T cells ($p < 0.001$) and only moderately elevated compared to normal CSF specimens ($p < 0.040$). In MS patients with progressive neurological symptoms, the percentages of CSF Fcγ cells and OKT_8+ cells were lower (6.40 ± 5.91, $p < 0.001$; 22.53 ± 10.87, $p < 0.012$) compared to controls. In contrast, the percentage of OKT_4+ cells (61.67 ± 12.87, $p < 0.011$) was higher. In patients with progressive MS, the percentage of PB B cells was found to be higher (23.98 ± 9.17) as compared to controls (18.33 ± 7.67, $p < 0.008$) and to MS patients with stable neurological symptoms (19.28 ± 8.16, $p < 0.036$). In these patients, the percentage of PB lymphocytes, carrying IgG (9.65 ± 5.03), was higher when compared to individuals with stable MS (6.65 ± 3.81, $p < 0.007$) and controls (6.00 ± 8.57, $p < 0.002$).

In MS patients with progressive neurological symptoms, a negative correlation was noticed between the percentage of CSF lymphocytes and the CSF OKT_4+/OKT_8+ cell ration ($p < 0.018$), and between the percentage of CSF lymphocytes and the percentage of PB lymphocytes stained with fluorescein conjugated antiserum to kappa light chains ($p < 0.006$). There was an indication of a positive correlation between the CSF IgG concentration and the PB OKT_4/OKT_8+ cell ration ($p < 0.044$).

In MS patients presenting with progressive neurological symptoms, the most frequent abnormalities found were a decreased percentage of Fcγ cells (86%) and a decreased OKT_8+/OKT_4+ cell ratio (75%) in the CSF. There was no correlation between the percentage of CSF Fcγ cells and the CSF OKT_8+/OKT_4+ cell ratio. In 41% of MS patients with progressive neurological symptoms, an increased percentage of CSF OKT_4+ cells was observed. A decreased percentage of CSF OKT_8+ cells was found in 32% of MS patients in this series.

In 14% of MS patients with progressive neurological symptoms, a decreased percentage of CSF Fcγ cells was found in individuals with a normal CSF OKT_8+/OKT_4+ cell ratio. In 9% of the MS patients with progressive symptoms, examination of CSF and PB for T cell subpopulations revealed abnormal findings only in the CSF. Considering MS patients with progressive neurological symptoms in whom only one of the PB immunologic parameters studied was abnormal (one patient with a decreased percentage of T cells, an increased percentage of OKT_4+ cells, and an increased percentage of OKT_3+ cells, and two patients with an increased percentage of B cells) and patients with progressive MS with normal PB studies, negative or inconclusive findings in PB were obtained in 20% of MS patients with progressive neurological symptoms in whom simultaneous CSF examination revealed definite abnormalities in the distribution of T cell subpopulations.

In 87 MS patients with repeated PB studies, marked changes in the percentages of PB, T, Fcγ, OKT_4+, OKT_8+, and B cells were noticed, not infrequently with nonsignificant changes in their objective neurological symptomatology. In 13 controls subjects, with a total of 37 PB examinations over a period of time extending up to two years, a decreased percentage of Fcγ cells was established on one examination (2.7%), a decreased OKT_8+/OKT_4+ cell ratio in two instances (5.4%) increased percentage of B cells in three instances (8.1%), and a decreased percentage of T cells on seven examinations (18.9%).

DISCUSSION

In MS patients with progressive objective or subjective neurological symptomatology, abnormal distribution of CSF T cell subsets may be expected in about 20% of individuals with nonrevealing PB studies. MS patients with progressive neurological symptoms may have a decreased percentage of Fcγ cells in CSF or PB in the absence of abnormalities in the distribution of OKT_4+, and/or OKT_8+ cells. In age and sex matched controls of our MS patients, transient decrease or increase in the percentages of PB T cells and their subpopulations and/or B cells were found in the absence of abnormal clinical signs. However, the incidence of abnormal findings in controls was lower as compared to MS patients with a decreased percentage of PB Fcγ cells being the least frequent and decreased percentage of rosetting T cells, the most frequent abnormalities. Abnormal percentages in PB Fcγ, OKT_4+, OKT_8+, and B cells may be found in MS patients complaining of progressive motor deficits, muscle spasticity or increasing fatigue without objective signs of worsening in neurological symptoms. Similarly, in two of four MS patients studied, who indicated progressive subjective neurological symptoms, abnormalities in the distribution of CSF T cell subpopulations were found. In agreement with Hauser et al.,[15] we were unable to observe an accumulation of OKT_8+ cells in CSF in MS patients showing decreased percentages of PB OKT_8+ cells.

In view of the reported predominance of OKT_4+ cells in perivascular infiltrates in active demyelinating brain lesions[19,20] the negative correlation between the percentage of CSF lymphocytes and the CSF OKT_4+/OKT_8+ cell ratio in individuals with progressive MS and an elevated CSF cell count and the higher incidence of an increased percentage of CSF OKT_4+ cells (41%) compared to PB (25%) in the MS patients studied, the abnormal proportion of the T cell subsets established is very similar to the OKT_4+, and OKT_8+ cell distribution in lungs and PB of individuals with sarcoidosis and high intensity alveolitis[21] and in skin lesions of the tuberculoid form of leprosy [22].

Pathogenetic mechanisms in the active phase of MS appear to be mediated predominantly by OKT_4+ cells crossing the blood/brain barrier and by the subsequently released lymphokines. The lymphokines possibly induce alteration of myelin and activate a restricted number of locally available B cell clones which produce predominantly IgG with resulting oligoclonal gammopathy manifested most noticeably in the CSF.

Because there are no indications of significant, persistent correlations between PB T cells, T cell subsets and B cells in MS patients on repeated longitudinal studies, one can postulate that in MS patients, the immunopathologic process, controlled by the extraneural immune system, and particularly the immunopathologic mechanisms inside the blood/brain barrier, present themselves in overlaping phases involving, in succession, immunoregulatory T cells, B cells and the secondary production of immunoglobulins.

SUMMARY

The active phase of multiple sclerosis is associated with abnormal proportion of T cell subpopulations, more frequently demonstrated in the cerebrospinal fluid as compared to peripheral blood. No significant correlations were found between T, Fcγ, OKT_4+, and OKT_8+ cells in cerebrospinal fluid and peripheral blood of multiple sclerosis patients. In progressive multiple sclerosis, a decreased percentage of cerebrospinal fluid and circulating Fcγ cells may be found in absence of abnormalities in the percentage of OKT_4+, and/or OKT_8+ cells. An understanding of the dynamic changes in the proportion of T cells, their subsets and B cells in cerebrospinal fluid and peripheral blood of multiple sclerosis patients is essential for optimal application of intensive immunoregulatory therapy in patients with active multiple sclerosis.

Acknowledgements

The authors thank Mrs. Linda Monk, Mrs. Joyce Hardwick, Mrs Diana Albright, Mrs. Shirley Finchum and Miss Leann Allison for technical assistance.

REFERENCES

1. R. P. Lisak, A. I. Levinson, B. Zweiman, and N. I. Abdou, T and B lymphocytes in multiple sclerosis, Clin.Exp.Immunol., 22:30 (1975).
2. H. J. Sagar and I. D. Allonby, Lymphocyte subpopulations in multiple sclerosis, J.neurol Sci., 43:133 (1979).

3. J. R. Huddlestone and M. B. A. Oldstone, T suppressor (T_G) lymphocytes fluctuate in parallel with changes in the clinical course of patients with multiple sclerosis, J.Immunol 123:1615 (1979).
4. E. L. Reinherz, H. L. Weiner, S. L. Hauser, J. A. Cohen, J. A. Distaso, and S. F. Schlossman, Loss of suppressor T cells in active multiple sclerosis, N.Engl.J.Med., 303:125 (1980).
5. M.-A. Bach, F. Phan-Dinh-Tuy, E. Tournier, L. Chatenoud, J.-F. Bach, C. Martin, and J.-D. Degos, Deficit of suppressor T cells in active multiple sclerosis, Lancet 2:1221 (1980).
6. J. Sandberg-Wollheim, Lymphocyte populations in the cerebrospinal fluid and peripheral blood of patients with multiple sclerosis and optic neuritis, Scan.J.Immunol. 17:575 (1983).
7. J. F. Oger, B. G. W. Arnason, S. H. Wray, and J. P. Kistler, A study of B and T cells in multiple sclerosis, Neurology 25:444 (1975).
8. S. L. Hauser, H. L. Weiner, and K. A. Ault, Clonally restricted B cells in peripheral blood of multiple sclerosis patients: Kappa/Lambda staining patterns, Ann.Neurol 11:408 (1981).
9. J. C. Allen, W. Sheremata, J. B. R. Cosgrove, K. Osterland, and M. Shea, Cerebrospinal fluid T and B lymphocyte kinetics related to exacerbations of multiple sclerosis, Neurology 26:579 (1976).
10. P. K. Coyle, B. R. Brooks, R. L. Hirsch, S. R. Cohen, P. O' Donnell, R. T. Johnson, and J. S. Wolinsky, Cerebrospinal-Fluid lymphocyte populations and immune complexes in active multiple sclerosis, Lancet 2:229 (1980).
11. H. S. Panitch and G. S. Francis, T-lymphocyte subsets in cerebrospinal fluid in multiple sclerosis, N.Eng.J.Med., 397:560 (1982).
12. J. Oger, J. P. Antel, A. Noronha, and B. G. W. Arnason, Changes in T-cell subpopulations in the cerebrospinal fluid of multiple sclerosis patients, Neurology 32:A148 (1982).
13. N. Cashman, C. Martin, J.-F Eizenbaum, J.-D. Degos and M.-A. Bach, Monoclonal antibody-defined immunoregulatory cells in multiple sclerosis cerebrospinal fluid, J.Clin.Invest., 70:387 (1982).
14. H. Weiner, S. L. Hauser, S. F. Schlossman, and E. L. Reinherz, Analysis of CSF mononuclear cells in multiple sclerosis using monoclonal antibodies: Correlation with disease activity and peripheral blood T-cell subset changes, Neurology 33:A165 (1982).
15. S. L. Hauser, E. L. Reinherz, C. J. Hoban, S. F. Schlossman and H. L. Weiner, CSF cells in multiple sclerosis, Monoclonal antibody analysis and relationship to peripheral blood T-cell subsets, Neurology 33:575 (1983).
16. S. L. Hauser, E. L. Reinherz, C. J. Hoban, S. F. Schlossman and H. L. Weiner, Immunoregulatory T-cells and lymphocytotoxic antibodies in active multiple sclerosis: Weekly analysis over a six-month period, Ann Neurol., 13:418 (1983).

17. J. R. Huddlestone and M. B. A. Oldstone, Suppressor T cells are activated in vivo in patients with multiple sclerosis coinciding with remission from attack, J.Immunol., 129:915 (1982).
18. O. J. Kolar, P. H. Rice, F. H. Jones, R. J. DeFalque, and J. Kincaid, Cerebrospinal fluid immunoelectrophoresis in multiple sclerosis, J.Neurol Sci., 47:221 (1980).
19. H. Nyland, R. Matre, S. Mørk, J.-R. Bjerke, and A. Naess, T-lymphocyte subpopulations in multiple-sclerosis lesions, N.Engl.J.Med., 307:1643 (1982).
20. C. J. J. Brinkman, H. H. ter Laak, O. R. Hommes, S. Poppema, and P. Delmotte, T-lymphocyte subpopulations in multiple sclerosis lesions, N.Engl.J.Med., 307:1644 (1982).
21. G. W Hunninghake and R. G. Crystal, A disorder mediated by excess helper T-lymphocyte activity at sites of disease activity, N.Engl.J.Med., 305:429 (1981).
22. W. C. Van Voorhis, G. Kaplan, E. N. Sarno, M. A. Horwitz, R. M. Steinman, W. R. Levis, N. Nogueira, L. S. Hair, C. R. Gattass, B. A. Arrick, and Z. A. Cohn, The cutaneous infiltrates of leprosy: Cellular characteristics and the predominant T-cell phenotypes, N.Engl.J.Med., 307:1593 (1982).

DISCUSSION

Raymond Roos

The University of Chicago, Department of Neurology
Chicago, Illinois, USA 60637.

Dr. Kolar has contrasted his results with those of other groups and indicated the differences in results obtained in different series. These differences are clearly of concern if one expects the results to influence treatment. Why is this variability present? First, one can wonder whether the monoclonal antibodies presently used as markers for lymphocyte subpopulations are the best ones, or whether they are used merely because they are the first ones. Second, one can question whether the study of cerebrospinal fluid (CSF) cells adequately reflects the status and activity of inflammatory cells within the central nervous system (CNS). Third, flow cytometry, rather than fluorescent microscopic enumeration of the cells, may be a preferable and more accurate means of analyzing the small numbers of CSF lymphocytes; it is also possible of course that flow cytometry may be measuring somewhat different parameters. Last, some of the clinical correlations of studies may be flawed because of difficulties in assessing disease activity; the availability in certain centers of nuclear magnetic resonance and position emission tompgraphy may improve our ability to determine disease activity.

Despite these controversies there is a general agreement concerning the presence of an increased number of T cells in the CSF and a decline in suppressor cells in blood and the CNS. Suppressor T cells may not only differ in number in MS but also may be _functionally_ different. The reason for the decline and changes in the suppressor T subpopulation is unclear. Several explanations have been suggested. (1) Suppressor T cells move from the blood into the CNS compartment. The decreased number of T cells in the CSF and their absence from the perimeter of the CNS parenchyma makes this hypothesis unlikely, although it is also possible that our autopsy CNS studies have not been idea. (2) Suppressor T cells are destroyed. (3) Suppressor T cells are immunologically modulated under the influence of anti-lymphocyte antibodies, which are known to occur in multiple sclerosis (MS). Antibody could strip the cell surface markers, reduce suppressor activity, but not destroy the cell. The absence of cytodestruction seen with lymphocyte antigenic modulation may make future prospects of monoclonal antibody immunotherapy difficult.

What is the importance of a decline in suppressor cells? The suppressor cell decline may merely reflect the immunodysregulatory state, i.e. the suppressor cells are nonsense T cells directed against antigens irrelevant to MS pathogenesis (or perhaps "missense" T cells is a better expression). Or perhaps the suppressor cell decline may permit the production of an important pathogenic antibody. Alternatively, the suppressor cell decrease could be the result of a pathologic antibody directed against white matter which happens to cross-react with suppressor cells. The availability of MS T cell clones may make studies of their antigenic target(s) possible.

IMMUNOLOGICAL TREATMENTS IN MULTIPLE SCLEROSIS: RATIONALE, RESULTS AND NEW AVENUES

R. E. Gonsette

Medical Director
Belgian National Centre for Multiple Sclerosis
Melsbroek, Belgium

INTRODUCTION

Cytotoxic drugs have been used in multiple sclerosis (MS) since 1966. At that time however, our understanding of immune mechanisms involved in the development of the disease was very limited.

At first, the rational of immunotherapy in MS was based on the existence of an increased antibody production in the central nervous system (CNS), revealed by the oligoclonal aspect of immunoglobulins G (IgG) in the cerebrospinal fluid (CSF). This humoral immune response appeared to be of importance in the pathogenesis of the disease, and the goal was therefore to abrogate this abnormal IgG's synthesis by killing the antibody producing B cells.

Since then, it appeared that most of the CSF antibodies and IgG production are non relevant phenomena that do not correlate either with the disease activity or with its pathogenesis, and that imbalance in cellular immunity is probably more important. From the other side, our knowledge of the effects of cytostatic or cytotoxic agents on the immune system is progressing and those substances are used more and more frequently as immunomodulators rather than as cytotoxic drugs. Moreover, non cytotoxic immunomodulators have been recently proposed even though their exact mechanisms of action remain to be defined.

Our concepts concerning the pathogenesis of MS are certainly less obscure than 10 years ago, and the first table summarizes what we know and what remains controversial concerning the putative causes and biological mechanisms involved in the disease progression.

Table 1. Possible Pathogenic Mechanisms in Multiple Sclerosis

What we know	What we do no know
Causes:	Causes:
External factor	Virus(es)?
Associated immune processes	Auto-immune disease?
Genetic background	Nature of the genetic predisposition?
Mechanisms:	Mechanisms:
Demyelination	Target: oligo and/or myelin sheath?
Immune process	Humoral and/or cellular?
Immune injury	Antibody + complement?
	Immune complexes?
	Cytotoxic T cells?
Immune imbalance	Antibody dependant cellular toxicity?
(H/S) associated to disease activity	Loss of S cells induces demyelination?
	Demyelination induces loss of S cells?
	Both induced by a third factor?

There are still more questions than answers, but we have to accept that abnormal immune reactions are associated with the evolution of the disease, particularly during active phases.

In spite of all those uncertainties, concerning both the immune basis of MS pathogenesis and the exact mode of action of immunosuppressors, considerable efforts have been made since 1966 to reduce or correct immune abnormalities observed in MS. The aim was of course to influence the inexorable evolution of the disease in the long run by reducing the relapse rate frequency and above all, the slowly progressing neurological and functional disability.

The first paper dealing with immunosuppression in MS: "Sclérose en Plaques et Processus d'Auto-Immunisation: Traitement par les Antimitotiques" was published in 1966 by Aimard et al., Cyclophosphamide was used but, because of its subjective intolerance, it was soon replaced by Azathioprine. Since this first attempt, many other clinical studies have been published using various substances, and from 1966 to 1977, nearly 800 MS patients have been treated.

The results of those efforts were summarized during the First Workshop on Immunosuppressive Treatments (April 1,2 - 1977 Melsbroek, Belgium). At that time, some beneficial effects were found in uncontrolled studies, and adverse reactions appeared acceptable. It was therefore decided to continue, using a methodology as strict as possible and particularly:

- controlled groups and when feasible, blind studies
- appreciation of the progression of the disease determined by usual disability and/or neurological scales
- large number of patients and a follow-up for at least 3-4 years
- treatment begun within the first 10 years of the evolution when a progression of the disease is still to be expected.

The purpose of this paper is to review the clinical trials published since 1977, using different techniques (Table 2). Some of them have been discussed during the Second Workshop on Immunosuppressive Treatments in MS (June 10,11,12 - 1982, Nijmegen, The Netherlands). Since then however, new progress has been made in neuroimmunology and within the framework of this meeting, we thought it interesting to review those therapeutic efforts again and for each treatment to state the rationale and the possible effects on the immune system, as well as the therapeutical scheme and the clinical results. To make the text as clear as possible, these data will be summarized in tabular form.

RECENT IMMUNE TREATMENTS IN MULTIPLE SCLEROSIS

Azathioprine

Azathioprine is certainly the most widely used substance. Its immune effects in laboratory experimental models are summarized in Table 3.

Clinical immune effects of Azathioprine in MS patients have been studied by Oger et al. (Table 4).

According to those observations, T cell suppressor activity is lower in MS patients when compared to healthy persons, and Azathioprine reduces this suppressor activity still further. On the other hand, the intrathecal production of IgG is restored to normal values.

Table 2. Immune Treatments Discussed in this Review (1977-1983).

Antilymphocyte Serum	Levamisole
Antithymocyte Globulin	Lymphocytapheresis
Azathioprine	Methotrexate
Cyclophosphamide	Myelin Basic Protein
Cyclosporine A	Plasmapheresis
Cytarabine	Polyunsaturated Fatty Acids
Hyperbaric Oxygen	Synthetic Polypeptide (COP I)
Immunoglobulin G	Transfer Factor
Interferon	Thymectomy

Table 3. Immune Effects of Azathioprine.

<u>Humoral Immunity</u>	
B cells:	marked depression (?)
	lowered IgG secretion
<u>Cellular Immunity</u>	
T cells:	?
H cells:	resistant
S cells:	marginally lowered (Oger 1982)
	markedly decreased (Trotter 1982)

NB. In conventional dosages corrects abnormal B cells without altering T cell regulation
At high doses S cells are affected

Table 4. Effect of Azathioprine on Suppressor Activity and IgG Production. (2 mg/kg/day) (Oger et al., 1982).

Group	Patients n =	IgG Production ng/ml	Patients n =	Con A induced suppressor activity (%)
Controls	32	1.377	44	40
MS stable	18	2.372	22	30
MS treated	12	1.238	10	23

Table 5. Azathioprine (Rosen et al., 1979).

127 patients			
Daily dose:	2 mg/kg		
Mean follow-up:	Pilot study: 9 years (n=85)		
	Randomized group: 4.5 years (n=42)		
Results:			
Pilot study:	"<10% of the patients progressed from ambulatory to immobile status"		
Randomized group:			
		<u>Aza</u>	<u>Untreated</u>
	Stable patients	91%	35%

Azathioprine appears therefore mostly effective on humoral immunity even though its depressive effect on B cells is not always accepted.

The results of Azathioprine chronic treatment in MS have been published by Rosen et al., in 1979 (Table 5).

According to this randomized study, Azathioprine seems to exert a favorable effect on the progression of the disease since stable patients are found more frequently in the treated than in the non-treated group.

The results of Aimard et al., have been recently published (1983) (Table 6).

The results of this controlled study with long-term administration of Azathioprine appear in favor of a beneficial effect of the drug on the progression as well as on the annual relapse rate (ARR).

The series of Sabouraud et al., has been published in 1982 (Table 7).

In this uncontrolled study, patients with a pure remittent form seem to benefit the most from Azathioprine since there is a striking

Table 6. Azathioprine (Aimard et al., 1983).

175 patients, mean follow-up 9 years
Daily dose: 150 mg
Control group: 128 patients
Results:

	Treated	Control
Stable	45%	13% (P 0.001)
Stable + bouts	36%	35% N.S.
Worsened	18%	53% (P.0.001)

Relapse rate (mean interval of time between bouts):

Before Aza:	2.5 years
After Aza:	3.8 years

Table 7. Azathioprine (Sabouraud et al., 1982).

67 patients, follow-up ⩾ 5 years
No control group
Daily dose: 3 mg/kg
Results: - no effect in
* remittent progressive forms (n = 17)
* pure progressive forms (n = 10)

- in 40 pure remittent forms

	Before treat.	After treat.
ARR	0.77	0.16

- global stabilization: 40/67 cases (60%)

Table 8. Azathioprine (Zeeberg et al., 1982).

22 patients:	12 Aza
	10 placebo
Follow-up:	18 months
Dose:	2.5 mg/kg/day
Results:	no difference between both groups

reduction of the ARR. In addition, the progression of the disease appears lower in 60% of the patients.

The first Azathioprine blinded study has been published by Zeeberg et al., in 1982 (Table 8).

No difference was found between both groups, but the number of patients is small and the follow-up rather short. An interesting observation in this study is that the progression of the disease was slower in both groups when compared to untreated patients. This of course may result from a placebo effect, but a placebo effect lasting for 18 months is unlikely. We rather think that this study demonstrates the necessity of a continuous follow-up of MS patients in order to prevent and/or to treat minor complications early (e.g. urinary bladder infections) and to avoid their influence on the progression of the disease.

The final report concerning a double blind trial of Azathioprine has been published by Mertin et al. (1982) (Table 9).

In fact, this study combines an acute immunosuppression (associating antilymphocyte globulin and prednisone) and a chronic administration of Azathioprine for 14 months.

No significant differences were found between the treated and the placebo groups, but the authors consider that there is a definite trend in the treated patients towards a slower evolution and a lower ARR.

The interval of time between the discontinuation of treatment and first relapse is longer in the treated group, and this is in favor of a potential efficacy of Azathioprine. The better response in female patients observed in the preliminary publication (1980) was not confirmed, and no conclusion could be drawn in this study about the role of HLA antigens in determining the response to the treatment.

Another controlled study has been published by Patzold et al., in 1982 (Table 10).

Table 9. Azathioprine (Mertin et al., 1982).

43 patients:	21 treated 22 placebo	
Acute treatment:		
- Antilymphocyte globulin:	500-750 mg daily, for 15 days	
- Prednisone:	150 mg/day, subsequently tapered (4 weeks)	
Chronic treatment:		
Azathioprine:	3 mg/kg/day, for 14 months	
Evolution:		
	Aza	Placebo
Stable	16	13
Worsened	5	9
Annual relapse rate:		
Before treatment	1	1
After treatment	0.7	1

Table 10. Azathioprine (Patzold et al., 1982).

115 patients:	60 treated 55 placebo	
Daily dose:	2 mg/kg	
Follow-up:	1 year n = 17 2 years n = 98	
Results (2 years follow-up):		
Mean progression ($\bar{x}$)		
	Aza	Controls
Intermittent progressive	15.10	28.37
Intermittent	9.33	10.95
Chronic progressive	27.96	12.76

In this randomized controlled study, in opposition to the results of Sabouraud et al., there is no clear influence of Azathioprine on the course of patients with a pure relapsing form, as well as with a pure progressive evolution.

In opposition, in the group of patients experiencing intermittent progressive course, there is a favorable effect since the mean progression is significantly lower in treated patients (P=0.01019). Of note is that the standard deviation is quite important in each group.

A recent clinical study has been published by Lauer et al. (1984) (Table 11).

This study emphasizes the difficulties inherent to randomized blinded studies since the course of the disease was more active in the treated group. When comparing similar periods of time before and after treatment however, a favorable incidence of Azathioprine on both the progression index and the ARR was observed.

Adverse effects of Azathioprine have been reviewed by Patzold et al. (1982) (Table 12).

In fact, no serious side-effects have been reported so far and the incidence of cancer in Azathioprine treated MS patients does not seem higher than in the general population.

Cyclophosphamide

Cyclophosphamide (CY) has been less widely used in MS because of its subjective intolerance. The immune response to CY is summarized in Table 13.

Table 11. Azathioprine (Lauer et al., 1984).

62 patients:	46 relapsing remittent 16 progressive
Control group:	59 matched patients (sex, onset, type, course)
Dose:	150 mg daily
Follow-up:	3-6 years
Results:	

Treated Patients	Entry	End	After Withdrawal
ARR	1	0.48	1
Progression index	0.76	0.16	0.58

Treated Patients vs controls	Aza	Untreated
ARR	0.48	0.32
Progression index	0.16	0.25

Bias: treated patients had a more active course
Comparison of similar pre- and post-treatment periods:

	Aza	Untreated	
Reduction ARR	0.60	0.23	(P<0.05)
Reduc.Progr.Index	0.57	0.32	(N.S., P>0.1)
% Pat. not progress.	67%	50%	(N.S., tendency: P<0.10)

Table 12. Adverse Effects of Azathioprine (Patzold et al., 1982)

Material:	105 cases
Adverse effects:	
- Haematological changes	2 severe anaemia
- Gastrointestinal complications:	3 major intolerances
- Hepatotoxicity:	increased liver enzymes:24% 2 severe hepatitis
- Infections:	seldom observed (acne)
- Allergic reactions:	2 cases
- Malignancies:	4 cases no proved causal relationship

Table 13. Immune Effects of Cyclophosphamide

Humoral Immunity	
B cells:	marked depression dose dependent long lasting superior to other alkal. agents
Cellular Immunity	
T cells:	depression
H cells:	highly vulnerable
S cells:	depression (Turk 1979) increase (Brinkman 1982)

NB: Changes are not due to cell death
Altered B and T cells functions are reversible
Peak CSF concentrations vs serum: parent subst. 50% alkal. fraction 20%

On experimental basis, CY influences both humoral and cellular immunity and this could be an advantage if both immune responses are involved. As a matter of fact, in a preliminary study of acute CY immunosuppression in 5 patients, we observed that T cells are more severly depressed than B cells, and H cells than S cells. According to Brinkman (1982), S cells would be even increased during chronic administration, with a simultaneous return of the H/S ratio normal values.

CY has been used by Hommes et al., in chronic progressive forms and his therapeutic scheme as well as his recent results are summarized in Table 14.

Progressive froms of MS are in theory the most reliable ones to appreciate the potential effect of treatments, since the evolution is slowly and inexorably progressing, without any stabilization nor improvement. Keeping that in mind, it appears that after a mean follow-up of 2.5 years, 43% of CY treated patients are stable. According to Hommes, this high percentage of stable patients does not correspond to the natural evolution of the progressive form of the disease. Of note is that Prednisone was systematically associated to CY in this series.

Since 1966, over 250 patients have been treated with CY at the Belgian National Center for MS in Melsbroek. The first protocol study (Gonsette et al., 1977) is summarized in Table 15.

Table 14. Cyclophosphamide (Hommes et al., 1982)

CY (400 mg) + Prednisone (100 mg) daily for about 3 weeks

Pure progressive form:	n = 47
Follow-up:	1 - 5 years (mean 2.5)
Results:	

	n =
Stable	10 (43%)
Worsened (DSS+1)	12 (25%)
DSS + 2	15 (32%)

Best results in patients with

- a higher progression rate
- an evolution shorter than 6 years
- HLA typing: DR W 2

Table 15. Cyclophosphamide (Gonsette et al., 1977).

Phase I: Methodology

Acute i.v. CY immunosuppression (mean dose 4.6 g)
Duration of treatment: 3 weeks
Leucopenia ≤ 2000 - Lymphopenia ≤ 1000
No cortisone or ACTH during or after CY
Relapsing remittent or relapsing progressive forms
GOAL: possible effect on the ARR
Evaulation:

- ARR calculated over a 2 years period before and after CY in the same patients
- ARR in treated patients vs retrospectively matched control group

The aim of this phase I study was to determine if a transient, acute, intensive immunosuppression was able to influence the ARR in patients with frequent exacerbations. Of note is that patients with a recent relapse (less than 8 weeks) were not included and that no corticoid therapy was associated. It was therefore to be expected that, if observed, clinical changes would be the results of CY therapy.

The results observed in 119 patients with a follow-up for 2 years and over are summarized in Table 16.

When patients are taken as their own controls, and when the ARR is calculated over a short period (2 years before and after treatment), a definite reduction of the ARR was observed after CY immunosuppression. This reduction was more pronounced in patients with a shorter evolution.

To overcome the objection of patients serving as their own controls, the ARR of treated patients was compared to that of untreated patients, retrospectively matched, according to the duration of the disease, the clinical form, the evolution, the sex and age (Table 17).

The mean spontaneous ARR decrease in a group of 91 non-treated patients, calculated over the 2 preceeding years is of 35% at the end of the 4th year, 18% at the end of the 6th year and 13% at the end of the 8th year. In the CY treated patients, the reduction of the ARR is definitely higher, even in patients with a pre-treatment evolution of 8 years.

However, in spite of its short term favorable influence on the ARR, CY acute immunosuppression was unable to change the progression rate of the disease and a second trial protocol associating acute and chronic CY immunosuppression was designed (Table 18).

Table 16. Cyclophosphamide (Gonsette et al., 1977).

<u>Phase I</u>: Results (Patients as their own controls)

Duration of disease before CY	Patients n =	Annual Relapse Rate Before CY	After CY
1 - 2	24	2.6	0.3
3 - 4	35	1	0.27
5 - 6	30	0.90	0.31
7 - 8	21	0.77	0.25
9 -10	19	0.46	0.46
Total	119		

Table 17. Cyclophosphamide (Gonsette et al., 1977).

Phase I: Results (Treated patients vs control group)
Percentage decrease of ARR vs 2 precedent years

Duration years+	Untreated patients n = 91 (spontaneous ARR decrease)	Treated patients n = 91
1-2	-	81%
3-4	35%	73%
5-6	18%	65%
7-8	13%	67%

+ Treated patients : duration of disease before CY
Untreated patients : duration of disease

Table 18. Cyclophosphamide (Gonsette et al., 1984)

Phase II: Methodology

Acute CY immunosuppression followed by a chronic treatment with cyclophosphamide 50 mg/day
or isophosphamide 150 mg/day

Patients

- experiencing frequent relapses (mean ARR 1.62)
- with a short evolution (mean duration 4 years)
- low disability status (Kurtzke $\leqslant 6$)

Evaulation:

DSS after $\geqslant 2$ years (mean follow-up 4 years) in treated patients and matched control group

The aim of phase II study was to evaluate the longterm influence of the treatment on the disability score. Chronically treated patients were compared to patients treated by a single acute CY immunosuppression and matched according to their age, sex, length of evolution, handicap at treatment, form of the disease, etc.. (Table 19).

After a mean follow-up for 4 years, 29/35 patients were stable in the chronically CY treated group compared to 12/35 in the control group. The difference between both groups is highly significant (P<0.005).

A study of Hauser et al. (1983) has been widely discussed in the recent literature. The results of this three arms clinical trial are summarized in Table 20.

Table 19. Cyclophosphamide (Gonsette et al., 1984)

Phase II: Results
Mean follow-up: 4 years (2-8 years)

Treatment	Patients n =	Stable n =	Worsened n =
Acute CY only	35	12	23
Acute + chronic CY	35	29	6

Table 20. Hauser et al., 1983

58 patients with severe relapsing and/or progressive forms

Treatment (2-3 weeks)	Patients N =	Stabilization DSS	Clinical	Aggravation Ambulant at entry Wheel-chair 1 year later
ACTH	20	20%	10%	11
CY + ACTH	20	80%	70%+	2
Plasm. Exch.	18	50%	44%	6

As a matter of fact, there is a strong tendency in this study in favor of CY treated patients when compared to ACTH or plasma exchange groups.

Unharmful side-effects of CY are well known: nausea, vomiting, hairloss, and discoloration of the nails. Non-specific adverse reactions are exceptionally observed: water intolerance, hyponatremia, drop in blood sugar in diabetic patients, prolonged apnoea after anesthesia and blurring of vision. Specific adverse reactions have been described, but their frequency is still controversial: carcinogenicity (reticulum cell sarcoma, squamous cell carcinoma, acute myeloid leukemia), organ toxicity (acute myopericarditis, hemorrhagic cystitis, acute hemorrhagic pancreatitis) and teratogenicity (azoospermia, amenorrhea, chromosome abnormalities, foetal death, extremity defects).

In our experience, side-effects of CY chronic administration mostly concern gastro-intestinal intolerance and their frequency is usually reduced by slowly increasing the dose (mean daily dose 50 mg). The cancer incidence in over 250 patients treated with CY since more than 15 years, is not higher than in a normal population.

Mechloretamine

Mechloretamine (Mustagen) is an analogue of CY. It has been

used in MS patients by Alexandrowicz et al. (1982) (Table 21).

The results of this preliminary study, associating an immunosuppressor (nitrogen mustard) and an immunomodulator (levamisole), encouraged the authors to design a multicenter longterm evaluation (10 years) with this treatment.

The results of an open clinical trial associating CY and Azathioprine have been published by Dachsel et al. (1982) (Table 22).

Table 21. Mechloretamine (Alexandrowicz et al., 1982).

50 patients with relapsing and/or progressive forms
"Mustagen" i.v. inj. for 5 days, monthly Levamisole 150 mg, twice a week

Duration of treatment: 2 years
Results:

	n =
Stable, no bouts	15 (30%)
Stable, less bouts	27 (50%)
Worsened	8 (16%)

Table 22. CY + Azathioprine (Dachsel et al., 1982).

82 patients:

- remittent progressive:	31
- secondary progressive:	39
- primary progressive:	12

Follow-up: 18 months

1. Intensive immunosuppression:
   - CY: 200-300 mg/day i.v. for 4-5 days
     100-150 mg/day (tablets) for 5 weeks
   - Simultaneously, Prednisone: 25 mg/day
2. Subsequent chronic immunosuppression:
   - Imuran: 100-150 mg/day for 18 months
   - Week-end drug holdiay

Results:

- Remittent progressive:	* no influence on ARR * definite beneficial effect on progressive deterioration
- Secondary progressive:	no influence
- Primary progressive:	markedly reduced progression

According to this open study, combined CY + Aza immunosuppression seems mostly effective during the active stages of the disease.

Methotrexate

Association of intramuscular and intrathecal injections of Methotrexate has been used by Dominguez et al. (1979).

There is no evaluation of the clinical results in this study but adverse effects are severe enough to preclude this particular treatment. Meningeal reactions are possibly due to the preservative substances added to Methotrexate solutions.

Cytarabine

The same holds for intravenous and intrathecal injections of Cytarabine as published by Tourtellotte et al. (1980) (Table 24).

Antilymphocyte Globulin or Serum

Longterm results of antithymocyte globulin (ATG) in MS have been published by Kastrukoff et al. (1978) (Table 25).

This open study failed to demonstrate a favorable influence of ATG therapy in the long run. Those results confirm therefore previous observations indicating that the beneficial effect observed in some patients immediately after antilymphocyte globulin therapy does not last very long.

Table 23. Methotrexate (Dominguez et al., 1979).

7 patients

Dosage:
- 10-12 mg i.m. per square meter body surface weekly for 6 weeks
- 1 intrathecal injection monthly for ?

Adverse effects:
visual alteration
delirium
aseptic meningitis

Results:
Clinical?

Immunological:	CSF IgG synthesis decrease Oligoclonal aspect disappeared in 1 case

Table 24. Cytarabine (Tourtellotte et al., 1980).

10 stable patients: 3 i.v
7 intrathecal

Dosage:
- i.v.: 67 mg per square meter body surface 3 times daily for 4-5 days
- intrathecal : 50 mg-1000 mg, 2-10 times

Follow-up:
2.5 years

Side-effects:
hemianopia
meningeal syndrome

Results:
Clinical: no improvement or worsening
Immunological: no change of CSF IgG synthesis oligoclonal aspect

Table 25. Antithymocyte globulin (Kastrukoff et al., 1978).

21 patients in relapse

Treated group (n = 10):
-ACTH: 40 IU/day i.v.
-ATG: 10 mg/kg/day i.v. for 28 days
or 20 mg/kg/day i.v. for 14 days

Control group (n = 11):
ACTH: 40 IU/day i.v.

Follow-up: 5 years

Results: no difference between both groups ($P > 0.5$)

Antilymphocyte serum (ALS) in association with Aza and Prednisone has been used in France by de Saxce et al. (1984) (Table 26).

de Saxce et al., conclude that Group I possibly yields better response than Group II, and that the difference is mostly marked during the first year. After discontinuation of ALS therapy, no significant difference could be found between both groups. This combined ALS and Azathioprine therapy does not seem therefore to influence the disease in the long run. Moreover, side-effects of antilymphocyte serum are not to be underestimated.

Table 26. Antilmyphocyte Serum (de Saxce et al., 1984).

Group I (n = 45)

ALS i.v. 125 mg active proteins/day for 2 weeks
3 times a week for 2 weeks
once a week for 1 year

Azathioprine 2 mg/Kg/day for 3 years
Prednisone 20 mg/day for 1 year

Group II (n = 22)

Same regime except no ALS

Results:		Group I	Group II
Follow-up		ALS + AZA + Pred.	Aza + Pred.
1st year	stable	76%	59%
	worsened	24%	41%
2nd year	stable	55%	33%
(stop ALS)	worsened	45%	67%
3rd year	stable	50%	24%
	worsened	50%	76%

Table 27. Thymectomy (Ferguson et al., 1983).

Group I (n = 16) : thymectomy + Azathioprine (1 year)
II (n = 16) : thymectomy alone
III (n = 32) : untreated control matched group

Follow-up: 2 years

Results (Groups I and II compared to Group III):
Group I: significant improvement (Kurtzke scale) more pronounced for relapsing remittent forms

Group II: no difference when compared to untreated patients

N.B.: The lack of side-effects of thymectomy in MS patients contrasts to usual experience in patients with myasthenia gravis

Thymectomy

The clinical results in thymectomy in MS have been recently published by Ferguson et al. (1983) (Table 27).

Apparently, association of thymectomy and Azathioprine yields favorable effects in relapsing remitting forms. However, even though thymectomy appears to be safer in MS than in myasthenia gravis, one may be reluctant to expose MS patients to a surgical procedure.

Lymphocytapheresis and Plasmapheresis

Removal of presumed cytotoxic or immune secreting lymphocytes is possibly obtained by lymphocytapheresis, and the results of Giordano et al. (1982) in MS are listed in Table 28.

Patients with relapsing or progressive forms seem to benefit the most from this treatment.

In their report, Giordano et al. (1982) briefly review the results obtained at other centers (Table 29).

Nearly 45% of the patients are reported to have shown improvement in these combined data. Of note is that in a blind study, lymphocytes were returned in 5 patients at the end of the procedure, and that 3 of them rapidly deteriorated afterwards. The length of the follow-up is not known and it is not possible to comment on the duration of the improvement.

Another short-term study of lymphocytapheresis has been published by McFarland et al. (1982) (Table 30).

Table 28. Lymphocytapheresis (Giordano et al., 1982).

70 out-patients

Lymphocytes yield per procedure 1×10^9-1×10^{10}
10 procedures over a 12 days period
When improved 1 session every 2 weeks for 3 months

Follow-up: 1 year

Results (Kurtzke DSS):

	n =	Before	After
Remitt. Relaps.	11	4.44	2.22 ($P < 0.005$)
Remitt. Progr.	45	6.02	5.48 ($P < 0.005$)
Chronic Progr.	14	7.92	7.64 ($P < 0.05$)

Table 29. Lymphocytapheresis: Data from a Multicenter Study (Giordano et al., 1982).

Authors	n =	Results (Kurtzke Scale) Improved	Stable
Hamburger et al.	6	4	2
De Green et al.	14	5	9
Kasprisin et al.	21	13	8
Pitts et al.	9	3	6

Table 30. Lymphocytapheresis (McFarland et al., 1982).

7 patients
Lymphocytes removed per procedure: 6 to 6 x 10^9
3 procedures a week for 3-4 weeks then weekly procedure for 3-4 weeks
Follow-up: 1-3 months

Results: patients		
	improved:	3
	unchanged:	2
	deteriorated:	2

Table 31. Lymphocytapheresis (Kennes et al., 1984)

Protocol I:
22 out-patients (chronic progressive)
1 x 10^9 lymphocytes yield
once a week for 3 weeks
once every 3 weeks for 6 months

Protocol II:
7 out-patients (chronic progressive)
once a week for 3 months

Follow-up: 9-12 months

Clinical results: modest but significant improvement in 29/29 patients

Immunological studies:
- Definite IgG serum level decrease (50%)
- T or B cells distribution unaffected
- H/S ratio normal before and after treatment

Cell-mediated immunity was studied in those patients by measuring *in vitro* response of lymphocytes to mitogen, and fluorescent activated cell scatter analysis was performed. No change of immunological functions could be observed.

A recent clinical and immunological study with lymphocytapheresis has been published by Kennes et al. (1984) (Table 31).

Up till now, available results do not suggest that lymphocytapheresis, an expensive procedure, carrying some risks has any utility in MS patients.

Rationale for plasmapheresis is summarized in Table 32.

Results of plasmapheresis in MS are summarized in Table 33.

Table 32. Rationale for Plasmapheresis.

Existence (?) in serum of MS patients of
- circulating immune complexes
- brain extracts complement fixing antibodies
- antimyelin antibodies
- antiligodendroglia antibodies
- demyelinating factor
- neuroelectric blocking activity factor

In 1980, 20 clinical groups have undertaken plasma exchange therapy in MS

Drawbacks: expensive and carries some risk

N.B.: concomitant Azathioprine and Prednisone administration

Table 33. Combined Data of Plasmapheresis in MS.

Year	Authors	Patients n =	Exchange n =	Results	
1980	Dau et al.	11	10-20	10/11	modest improvement
1980	Weiner et al.	8	3-5	6/8	modest improvement for 6 months
1981	Valbonesi et al.	11	5	4/6	moderate improvement
1982	Stefoski et al.	7	5	2/7	improvement
Immunological studies					
1984	Caparelli et al:	marked decrease of IgG (serum + CSF) CSF oligoclonal aspect unchanged			
1984	Guarneri et al:	significant decrease of polymorphonuclear leucocytes proteinase activity in active cases			
Neurophysiological studies					
1981	Rosen et al:	reduction in visual evoked potential latencies			
1982	Stefoski et al:	decrease of serum neuroelectric blocking activity			

It has been suggested that plasmapheresis, in association with immunosuppression drugs could improve our ability to modulate the immune system but no clinical studies supporting or denying its clinical utility in MS are available so far.

Levamisole

Levamisole is considered as an "immunomodulator" and its effects on the immune system are dose dependent (Table 34).

Usual therapeutical levels of Levamisole are considered as producing a decrease in activity of B cells and an increase of suppressor activity.

Levamisole has been used in MS since 1976 and the clinical results are summarized in Table 35.

In most studies, no definite effect of Levamisole on MS could be demonstrated and, moreover, it has been suggested by Dau et al., that this substance could be harmful. In a double blind study concerning 54 patients followed for 2 years, Gonsette et al., did not observe adverse reactions and it appeared that there was a tendency for treated patients to experience a slower progression of the disease and a lower ARR. The difference between the placebo and treated groups however is not significant. Most of the patients had a long evolution and a severe handicap at entry in the trial and it could be that Levamisole is more effective when applied earlier during the first years of the disease. Levamisole has been used with possible

Table 34. Immune Effects of Levamisole.

1. Low doses (potentiation)
   Humoral Immunity:
   - B cells: activated

   Cellular Immunity:
   - T cells: activated
   - H cells: marked increased
   - S cells: mild increased

   N.B.: increases heldper to a much important degree than suppressor activity
   net effect: immunopotentiation

2. High doses (suppression)
   Humoral Immunity:
   - B cells: not directly suppressed antibody synthesis reduced as a consequence of increased suppressor activity

   Cellular Immunity:
   - T cells: activated
   - H cells: ?
   - S cells: probably increased

   N.B.: marked increase of suppressor activity

Table 35. Clinical Trials with Levamisole.

Year	Authors	Patients n =	Follow-up	Results
1976	Dau et al.	7	3 months	5 patients deteriorate
1977	Myers et al.	4	1 year	no effect
1977	Cendrowski et al.	17	30 days	no effect
1977	Camenga et al.+	33	6 months	no difference treat./plac.
1979	Patzold et al.	23	10 months	no effect
1981	Massaro et al.	26	1 year	no conclusion
1982	Fonsette et al.+	54	2 years	Levamisole group more stable

+ Blind control study

benefit as a chronic immunomodulator after acute nitrogen mustard treatment (Alexandrowicz et al., 1982).

Hyperbaric Oxygen

Hyperbaric oxygen is able to reduce CNS lesions in EAR (Warren et al., 1978) and has been used therefore since as early as 1970 (Table 36).

From those preliminary open trials, it seems that HBO is able to provoke a mild and transient improvement in MS.

Longterm results of HBO have been published by Fisher et al. (1983) in a blinded randomized study (Table 37).

In fact, it is still difficult to comment on longterm benefit of this treatment, even though there seems to be a tendency for treated

Table 36. Combined data concerning HBO clinical trials.

Year	Authors	n =	Improvement
1970	Boschetty et al.	26	transient in 16
1978	Baixe et al.	11	moderate in 11
1980	Neubauer et al.	600	dramatic (39%) moderate (52%)
1980	Formai	?	positive results
1980	Pallotta	6	substantial in 6

N.B. no control group

Table 37. Longterm Results of HBO Therapy (Fisher et al., 1983).

40 patients: 20 MS (3 drop out)
20 controls

2 athmosphere absolute for 90 minutes daily for 20 days

Longterm results (1 year):

	Treated n = 17	Placebo n = 20
Patients improved	5	-
stable	10	9
worse	2 (12%)	11 (55%)

patients towards a slower progression. For this reason, a 2 years multicenter double blind control study has been designed in order to more definitely establish the value of HBO therapy in MS (Kindwall et al., 1983).

Transfer Factor

The exact nature and mechanisms of action of transfer factor (TF) are not known yet, but this substance is supposed to restore deficient cell-mediated immune functions. Several blinded studies have been published with conflicting results (Tables 38 to 41).

Behan et al. (1976) observed no evident difference between treated patients and the control group, and the same holds true for Fog et al. (1978) as well as for Collins et al. (1978).

Table 38. Transfer Factor (Behan et al. 1976).

29 patients:	14 transfer factor 15 placebo
TF:	isolated from one pint of venous blood Random donors
Dosage:	once a month for 3 months same procedure 3 months later
Follow-up:	1 year
Results:	

	TF	Placebo
Improved	3	2
Stable	8	9
Worsening	3	4

Table 39. Transfer Factor (Fog et al., 1978).

32 patients:	16 transfer factor 16 placebo		
TF:	1 unit = 1 - 2 x 10^7 leucocytes Random donors		
Dose:	3 units monthly for 13 months		
Follow-up:	1 year		
Results:		Neurological Examination	
		TF	Placebo
	Stable	23	25
	Worse	9	7

Table 40. Transfer Factor (Collins et al., 1978).

48 patients:	28 transfer factor 20 placebo		
TF:	2 x 10^7 lymphocytes/ml Random donors		
Dose:	10 ml every 2 months		
Follow-up:	1 year		
Results:		Neurological	Evaluation
		TF	Placebo
	Improved	3	1
	Stable	13	11
	Worse	12	8
Conclusions:	Results disappointing No effect in a number of serological and immunological studies Is the preparation biologically active?		

Another blind study has been recently published by Lamoureux et al. (1983).

During a follow-up period of 1 year, there is no clear influence of this short-term (3 months) treatment with TF.

A longterm study has been published by Basten et al. (1980) with more encouraging results (Table 42).

Table 41. Transfer Factor (Lamoureux et al., 1983).

27 MS patients:	14 transfer factor 13 placebo
TF:	2×10^7 lymphocytes/ml Unrelated donors
Dose:	2 ml subcutaneously at weekly interval Treatment period: 3 months
Immunological Studies:	TF do no stimulate the humoral response Significant decrease of total CSF protein IgG levels

In this prospective double-blind trial, the benefit of treatment with TF was not apparent until 18 months to 2 years. This could explain the failure of previous short-term studies to demonstrate any favorable effect. Another explanation may be that in other studies, random donors were used as the source of leucocytes for TF production. Even though statistical analyses are in favor of F therapy, it must be stressed that standard deviations in disability scores evaluation are relatively high. A carefully designed longterm study is underway at the Belgian Centre for MS (Van Haver et al., 1984) in order to demonstrate a possible differential effect of TF of MS relatives or random donors, and to establish the possible influence of a prolonged administration of this substance.

Table 42. Transfer Factor (Basten et al., 1980).

51 patients:	26 transfer factor 25 placebo
TF:	1 unit = 4×10^8 cells equivalents (mononuclear) Relatives as donors
Dose:	every 2 weeks for 1 month every 4 weeks for 6 months every 8 weeks thereafter Total: 25 doses
Immunological studies:	no clear cut effect
Results:	Percentage of Patients without Clinical Signs of Progression

	6 mo	12 mo	18 mo	24 mo
Treated	58	46	37	48
Placebo	56	36	16	24

Desensitization Techniques

Desensitization against a hypothetical encephalitogenic determinant has been tried in MS using myelin basic protein (MBP) (Table 43).

According to Campbell's and our own experience, no influence of MBP treatment in MS could be demonstrated. More recently, Romine et al. (1983) have used definitely higher doses without any therapeutic effect as well. Of note is that the development of a primary take by some patients in this study suggests that they had not been sensitized to MBP previously. Moreover, high doses of MBP may be associated with neurological adverse reactions. In neurophysiological studies, Seil et al. (1983) have shown that there is no correlation between serum demyelinating activity and the presence or absence of MBP therapy in MS patients.

Desensitization with a non-encephalitogenic synthetic polypeptide (COP-I) has been used by Bornstein et al. (1982) (Table 44).

Table 43. Combined Data of MBP Treatments in MS.

1973 Campbell et al.,	30	5 mg/week	some amelioration
1977 Gonsette et al.,	35	5 mg/week	no therapeutic effect
1983 Romine et al.,	19	75 mg/daily	no therapeutic effect

NB: In Romine's study, hypersensitivity to MBP was observed in 17/19 patients
* 15: delayed type (injection site)
  + neurological changes
    6 reversible
    2 residual
* 2: immediate type

Table 44. Synthetic Polypeptide -COP-I (Bornstein et al., 1982).

Open study:	16 patients (mostly chronic progressive)
Daily dose:	20 mg i.m.
Follow-up:	18 months
Results:	no effect 11/16 improvement 6/16

The results of this open study being in favor of a possible therapeutic effect of COP-I, a preliminary blinded trial was developed in exacerbating-remitting patients. According to a first analysis of the results, still blinded for the investigator, an external advisory committee has recommanded to extend the program to a full clinical trial in 1984.

Polyunsaturated Fatty Acids

The longterm results of a trial with polyunsaturated fatty acids (PUFA) have been recently published by Dworkin et al. (1984). Fifty-eight patients were included in a total series of 172 patients from other centers in UK and Canada (Table 45).

Treatment with PUFA significantly reduced the severity and duration of relapses and there was a tendency to a slower progression rate. Patients with remitting forms and with the shorter duration of illness appear to benefit more.

Table 45. Polyunsaturated Fatty Acids (Dworkin et al., 1984).

Blind Study:	58 patients:	29 linoleic acid 29 oleic acid	
Daily dose:	23 g		
Follow-up:	2 years		
Results:		Treated	Control
Kurtzke scale			
	Improved	8 } 21	7 } 18
	Stable	13	11
	Worsened	8	10
Millar scale			
	Score per relapse	8.2	18.2
Combined study:	172 patients:	87 treated patients 85 controls	
Follow-up:	28 months		
Results:		Treated	Control
Kurtzke DSS	Before	3.33	3.15
	After	3.79	3.93
Mean relapse score		15.29	26.47
Annual relapse rate		0.66	0.61

Interferons

Besides its well-known effect on the K-NK system, other influences of Interferons (IFN) on immune system are not clear yet (Table 46).

IFN production by peripheral blood mononuclear cells in MS patients has recently attracted attention (Table 47).

According to the most recent studies, IFN responders are less frequent among MS patients, but in those responders, IFN production appears normal.

Table 46. Immune Effects of Interferons

Humoral Immunity		
B cells:	antibody synthesis decreased	
Cellular Immunity		
T cells:	?	
H cells:	?	
S cells:	increased	
N.B.:	- NK system - cytotoxic cells - macrophages	activated

Table 47. Production of IFN in MS patients (Peripheral Blood. Mononuclear cells)

Year	Authors	Type of IFN	IN yield
1979	Neighbour	α	significant decrease
1980	Benczur	α,β	id.
1981	Neighbour	α,β,γ	id.
1981	Santoli	α	no difference
1982	Salonen	α	decr.: mumps-meales no diff. : Herpes simplex
1982	Zander+	α,β,γ	no difference
1983	Kaudewitz+	γ	id.
1983	Vervliet++	α,γ	id.
1983	Tovel	α	id.

\+ Design study excluding pre-existing genetic bias
++ IFN γ Responders: MS 37%, controls 85%

Table 48. Interferons (Clinical Trials).

1979 Ververken et al. (3 patients):
- i.m. 0.05 x 10^6 units IFN β /Kg body weight every other day for 12 days
- Follow-up: 9-18 months
- Results: no effect

1980 Fog et al. (6 patients):[2]
- i.m. HUIFN α 5 x 10^6 units daily for 2 weeks
  2.5 x 10^6 units daily for 5-13 months
- Follow-up: 1 year
- Results: no effect

1982 Cook et al. (9 patients):
- i.m. HUIFN β 1 10^6 daily for 2 months
- Follow-up: "over years"
- Results: "distinct improvement in some patients but not all" no firm conclusion

N.B.: general decrease in measles viral protein in the jejunum

Table 49. Intrathecal Interferon in MS (Jacobs et al., 1982).

- 20 patients: 10 MS
  10 controls
- intrathecal HUIFN β 1 x 10^6 IRU body square meter semiweekly for 4 weeks once a month for the next 5 months
- Mean follow-up: 1.9 years
- Results:

	Improved	Stable	Worse
IFN	5	3	2
Untreated	2	4	4

ARR (2 y):

	Before	After
IFN	1.8	0.2
Untreated	0.8	0.7

Clinical trials with IFN have been published by Ververken et al. (1979), Fog et al. (1980) and Cook et al. (1982), with disappointing results (Table 48).

The clinical study of Jacobs et al. (1982) using intrathecal injections of IFN seems more encouraging but has raised many reactions in the literature (Table 49).

The most impressive results concern the reduction of the ARR, but it has been objected that both groups of patients are small and not homogeneous concerning relapse rate, clinical form and progression of the disease.

Ruutiainen et al. (1983) have reported their experience with intrathecal injection of interferon α (Table 50).

No conclusion could be drawn on the therapeutic results owing to the limited number of patients as well as the restricted observation period.

In order to assess more definitely the therapeutic value of IFN in MS, several double-blind, randomized, cross-over studies are presently underway, but results will not be published until 2 or 3 years. A preliminary evaluation of the Californian study seems to indicate that IFN treated patients experience less attacks, but that the number of patients going into the progressive stage is the same in both groups. This apparent lack of influence on the progression of the disease is possibly due to IFN ability to activate macrophages (Johnson et al., 1983).

Immunoglobulin G

Clinical studies with immunoglobulin G (IgG) have been recently published by Schuller et al. (1983) and Guenther et al. (1984). Administration of exogenous IgG would produce an immunosuppression by a subsitution and a reduction of endogenous IgG synthesis. Moreover, IgG would induce a marked activation of T suppressor cells (Table 51).

After a 4 years follow-up, a slowing of the progression was observed by Schuller et al., in 2/3 of the patients and an objective improvement in 1/3. The study of Guenther et al., is more limited in

Table 50. Intrathecal IFN in MS (Ruutiainen et al., 1983).

- 5 severely disabled patients chronic progressive forms	
- 4 weekly intrathecal injections	
- Increasing doses:	0.1, 0.3, 1, 3 x 10^6 HUIFN α
- Follow-up:	3 months
- Side-effects:	Dose dependent (> 0.3 IU) Fever in all patients Extreme fatigue

time but again, a significant decrease of the disability score was noted after 1 year therapy. On the other hand, the modest reduction of the ARR must be viewed with caution.

THE PRESENT AND THE FUTURE

Attempts to influence the immune system in order to reduce the progression of MS disease fall into various categories: immunosuppression with cytostatic drugs, selective action on lymphocytes, possible immunomodulation, desensitization, plasma "toxic factors" extraction, and finally somewhat empirical techniques.

Among immunosuppressive substances, Aza and CY are mostly used. Usually, an acute intensive immunosuppression is achieved with high doses of CY, whereas Aza is chronically administered for several years. It must be reminded that high doses of cytostatic drugs alter lymphocytes viability, whereas chronic low doses exert a reversible influence on their functions.

During the last four years, 8 clinical studies using Aza in 524 patients have been reported. With the exception of 2 open studies, all clinical trials were blinded, including matched control groups. No significant differences were found between treated and untreated patients when the follow-up period did not exceed 2 years. In patients with a longer follow-up period did not exceed 2 years. In patients with a longer follow-up (3-9 years), a significant reduction of the progression index as well as the ARR was observed. Patients experiencing active forms of the disease seemed to benefit the most. However, due to the limited number of patients in each trial and to important standard deviations in some studies, those results must be interpreted rather as a significant tendency than as a definite evidence.

According to the recent literature, about 350 patients have been treated with CY. The subjective side-effects of this drug make blinded studies quite impossible and CY immunosuppressed patients are usually compared with matched untreated patients or with patients receiving other treatments. As previously mentioned, CY is generally used as an acute (pulse) intensive immunosuppression for 3 to 4 weeks, associated or not with Prednisone therapy.

It appears that after short term CY immunosuppression, a significant tendency towards a lower ARR and a slower evolution is observed in about 70% of the patients. This beneficial effect is mostly marked in patients with a definite progression or with a high ARR. However, this influence on the disease is transient, lasting for 1 to 2 years. For this reason, in recent studies chronic immunosuppression with CY, Levamisole or Aza was used after the acute treatment. Compared the transient effects of a single acute CY immunosuppression, it appears from those preliminary studies that, after a mean follow-up for 4 years, a combined acute and chronic CY therapy produced a significant reduction of the progression rate of the disease.

Immune treatments more specifically directed against lymphocyte cells (ALS, ALG, thymectomy, lymphocytapheresis) seem to yield only moderate and transient improvement, if any.

Plasmapheresis appears of marginal benefit in spite of its favorable effects on serum and CSF IgG production, visual evoked potential conduction, and serum neuroelectric blocking activity.

No clinical studies are available to date, definitely supporting or denying the clinical utility either of immunomodulators (Levamisole, Transfer factor, Interferons) or desensitization techniques (Basic Protein, Copolymer I).

The mechanisms by which hyperbaric oxygen, linoleic acid and Immunoglobulin G possibly influence the immune system remain unclear. Longterm linoleic acid supplementation reportedly reduces the severity and length of the relapses, without any evident influence on the progression rate. The utility of HBO is strongly debated and will only be established by the results of a multicenter clinical study presently underway. In the same way the potential improvement observed after infusion of high doses of IgG must be confirmed in controlled studies.

An analysis of the preceeding paragraphs provides definite arguments in favor of a real, even though moderate, effect of some immune treatments on the natural history of MS disease.

Prolonged immunodepression with cytostatic agents appears more effective than other techniques. CY, an agent acting on both humoral and cellular immunity, seems more potent than Aza, a drug mainly depressing the humoral effector arm.

Unfortunately, alkylating agents (CY, chlorambucil) are more carcinogenetic than antimetabolites (Aza). It is admitted that prolonged use of chlorambucil for non malignant diseases must be discouraged, and so far, 8 cases of acute leukemia have been reported in MS patients chronically treated with this alkylating agent.

Acute leukemia has been documented in some patients after prolonged administration of CY for non-neoplastic diseases. However, as far as we know, no malignancies related to CY immunosuppression in MS patients have been described.

For those two alkylating agents, the risk seems dose related and no acute leukemia has been reported after administration of a total dose smaller than 1 g for chlorambucil and 50 g for CY and urinary bladder cancer induced by CY has been observed after prolonged treatment with high doses (over 170 g for 4 years).

Acute myelogenous leukemia in patients with non-malignant diseases maintained on chronic immunosuppression with Aza, has been less frequently observed, and to date, no cases have been reported in Aza immunosuppressed MS patients.

In our experience, the prevalence of cancer for MS is the same as for the whole general population. In a series of 3000 patients, 21 certain cases of malignant disease were found (0,7%). Of note is that the prevalence of acute leukemia in this group (3 cases = 0,1%) is 10 times that of the general population (Seidenfeld et al., 1976). The same holds for rheumatoid arthritis, another immune disease (Isomaki et al., 1975).

Even though longterm complications of bladder cancer, leukemia or other teratogenic effects have not been reported after maintained immunosuppression in MS (with the exception of chlorambucil), they remain potential severe side-effects that must not be underestimated. Intermittent, repeated chemotherapy, or limiting treatment to as short duration as possible according to the evolution of the disease, may reduce the incidence of severe adverse reactions. However, we still need reliable parameters for an objective assessment of the biopathological status of MS patients. There is a reasonable hope that longitudinal monitoring of new immune parameters will help to delineate the real effects of immunosuppression as well as the fluctuations of the disease. Preliminary observations presented during this meeting by Kolar seem very promising to this effect.

The principle of immune treatments in MS is a valid one but we have to investigate new avenues in immunological manipulations to make them more effective at lower risks. It has been suggested that some actual treatments could be improved. Where EAE prevention by MBP is concerned, it has been shown that the efficacy is enhanced when administered together with Galactocerebroside. According to Raine (1983), MBP should act through a T cell response and Galactocerebroside through an antibody dependant response. In the same way desensitization in MS should be improved by addition of a lipid hapten to MBP.

Immunological effect of immunosuppressive agents have been recently studied by Shih et al. (1983) with a view to their possible therapeutical application in MS (Table 52).

According to the authors, CCNU presents several advantages including reduction of IgG synthesis, enhancement of suppressor activity, andy easy access to the CNS. As far as we know, no clinical studies have been performed with this agent in MS.

Table 51. Immunoglobulin G in MS.

Schuller et al. (1983)
- 31 patients
- i.m. injection of 5 g gammaglobulin, 3 times a week or i.v. injection, once a week
- Results:

Improved	11	20
Stable	9	
Worsened	11	

Guenther et al. (1984)
- 20 patients
- 5 g i.v. at 2 months interval for 12 months
- Results:

	Fog scale	Annual R.R.
Before treat.	28.8	1.13
After 6 months	21.3	1.02
After 1 year	19	0.99

Table 52. Immunological Effects of 3 Immunosuppressive Agents (modified from Shih et al., : Cl.Exp.Imm. 1983).

Parameters	Cyclophosphamide	CCNU	5-FU
T cells	augmented	no effect	decreased
B cells	decreased	no effect	no effect
S cells (FcR)	no effect	augmented	no effect
K activity	no effect	augmented	no effect
NK activiry	decreased	no effect	decreased
IgG synthesis	no effect	decreased	augmented

Cyclosporine A (CS-A) belongs to a new generation of immunocompetent agents and it has some potential advantages on previous substances. Its possible effects on the immune system are listed in Table 53. (Klaus, 1981).

An important advantage of CS-A should be its lack of antimitotic and cytostatic effects, but more experimental studies are needed.

The suppressive effect of CS-A on EAE has been studied as early as 1977 (Levine and Sowinski) and CY A is as effective as CY. Since there are strong arguments in favor of an influence of CY in MS patients, CS-A could possibly be beneficial too at lower risk. However prolonged administration of CS-A seems limited by its toxic effects on kidney and liver. Double blind studies are presently

Table 53. Immune Effects of Cyclosporine A.

<u>Humoral Immunity</u>	
B cells:	not directly suppressed antibody synthesis reduced as a consequence of H cells depression
<u>Cellular Immunity</u>	
T cells:	High affinity for proliferating T cells
H cells:	markedly impaired
S cells:	depressed (Borel 1980) mild induction (Routhier 1980)
N.B.:	Interferes with the inductive phase Specifically inhibits production of interleukines Acts selectively on immunocompetent cells Lacks antimitotic and cytostatic effects

under way in two neurological centers, but no results will be available before 1986.

Monoclonal antibodies (Mab) are wonderful biological tools not only because of their ability to define single antigenic determinants but also because of their potential capacity to modulate antigens specifically. Since our knowledge of immunoregulation is progressing faster and faster, the answer for the treatment of MS might be found in the manipulation of the immune system by idiotypic or anti-idiotypic antibodies directed at the clones of immunoreactive cells implicated in the pathogenesis of the disease. Of note is that Brinkman et al. (1984) have already been able to suppress EAE in Lewis rats by Mab to all peripheral T lymphocytes (W3/13).

CONCLUSION

The pressure to find relief for MS has never been so strong and we are indeed at a critical point in the search for a treatment.

In the beginning, immunosuppressive treatments were based more on hope and challenge than on clear scientific basis. Since many years most scientists are convinced that immune reactions are associated with the pathogenesis of Ms disease, but until recently, we were ignorant of the involved mechanisms.

During the last few years, our understanding of immune effector arms is advancing fast. Where MS is concerned, several observations suggest that the immune mechanisms involved in the progression of the disease are not necessarily restricted to the CNS but are possibly shared with other systems such as the lympoid tissues. It is striking to observe that MS lesions are preferentially localized around capillaries and ventricular walls, where peripheral immunocompetent cells come into contact with the CNS.

The exact nature of a putative interaction between CNS and extracerebral tissue remains unclear. One possibility is that MS is a "lymphoid-CNS" disease, which means that the immunizing event initiating the production of demyelinating factors occurs in extraneural lymphoid system years earlier, and that CNS demyelination represents a delayed reaction secondary to an autoregulation defect in lymphoid tissues. Another possibility is that specific antibodies produced in the brain enter the peripheral blood and recruit peripheral lymphocytes to migrate in CNS and bring about an antibody-dependant cellular cytotoxicity reaction against glial cells.

There are more and more arguments in favor of a peripheral origin of the lymphoid cells associated in the CNS with the development of MS lesions. Those observations circumvent the presence of the blood brain barrier and encourage attempts to influence the evolution of MS disease by manipulations of the immune system.

The major problem in evaluating the results of tenative treatments in MS is the impredictable evolution of the disease. There is little doubt however that this difficulty will be partly solved by a longitudinal monitoring of biopathological events with the recently discovered immunological tests and by appreciating the evolution of brain lesions by the nuclear magnetic resonance technique.

It is therefore to be expected that an objective appreciation of the efficacy of immune treatments in MS will be possible within the next five years. In the meantime it would be most unfortunate if untimely conclusions or inflated publicity about the results brought discredit on those clinical trials.

REFERENCES

Aimard, G., Girard, P. F., and Raveau, J., 1966, Sclérose en plaques et processus d' autoimmunisation. Traitement par les antimitotiques, lyon Med., 215:345.

Aimard, G., Confavreux, C., Ventre, J. J., Guillot, M., and Devic, M., 1983, Etude de 213 cas de sclérose en plaques traités par l'azathiorpine de 1967 á 1982, Rev.Neurol., 139:509.

Aleksandrowicz, J., Skotnicki, A. B., and Zduńczyk, A., 1982, The combined immunomodulatory treatment in patients with multiple sclerosis - A long-term study, Personal Communication.

Baixe, J. H., 1978, Bilan de onze années d'activité en médecine hyperbare, Med.Aer.Spatiale Med., Subaquatique Hyperbare, 17:90.

Basten, A., Pollard, J. D., Stewart, G. J., Frith, J. A., McLeod, J. G., Walsh, J. C., Garrick, R., and Van Der Brink, C. M., 1980, Transfer factor in treatment of multiple sclerosis, The Lancet, II:931.

Behan, P. O., Durward, W. F., Melville, I. D., McGeorge, A. P., and Behan, W. M. H., 1976, Transfer factor therapy in multiple sclerosis, The Lancet, I:988

Benczur, M., Petranyi, G. Gy., Palffy, Gy., Varga, M., Talas, M., Kotsy, B., Földes, I., and Hollan, S. R., 1980, Dysfunction of natural killer cells in multiple sclerosis: a possible pathogenetic factor, Clin.Exp.Immunol., 39:657.

Bolton, C., Borel, J. F., Cuzner, M. L., Davison, A. N., and Turner, A. M., 1982, Immunosuppression by cyclosporin A of experimental allergic encephalomyelitis, J.of the Neurol.Soc., 56:147.

Borel, J. F., 1980, Essentials of cyclosporin A. A novel type of antilymphocyte agent, Trends Pharmacol.Sci., 1:146.

Bornstein, M. B., Miller, A. I., Teitelbaum, D., Arnon, R., and Sela, M., 1982, Multiple Sclerosis: Trial of a synthetic polypeptide, Ann.Neurol., 11:317.

Boschetty, V., and Cernoch, J., 1970, Aplikace kysliku za pretlaku u nekterych Neurologiekych onemochemy, Bratisl.Lek.Listy, 52:298.

Brinkman, C. J. J., and Homes, O. R., 1982, The effect of cyclophosphamide on T lymphocyte and T lymphocyte subsets in patients with chronic progressive multiple sclerosis, in: "Immunosuppression in Multiple Sclerosis, Proceedings of the Nijmegen Workshop June 10-12, 1982. Clinical trials Journal 1984". O. Hommes, J. Mertin, W. W. Tourtellotte, eds., (to be published).

Camenga, D. L., 1977, Safety and efficacy of Levamisole in multiple sclerosis, in: "Excerpta Medica Proceedings of the 11th World Congress of Neurology, Amsterdam".

Campbell, B., Vogel, P. J., Fisher, E., and Lorenz, R., 1973, Myelin basic protein administration in multiple sclerosis, Arch.of Neurol., 29:10.

Capparelli, R., Inzitari, D., Sita, D., Guarnieri, B. M., Fratiglioni, L., Amaducci, L., Avanzi, G., Franco, C., and Lombardo, R., 1984, Plasmapheresis in multiple sclerosis: cerebrospinal fluid changes after treatment, in: "Immunological and Clinical Aspects of Multiple Sclerosis," R. E. Gonsette, P. Delmotte, eds, MTP Press Limited, Lancaster.

Cendrowski, W., and Czlonkowska, A., 1978, Levamisole in multiple sclerosis, with special reference to immunological parameters, Acta.Neurol.Scand., 57:354.

Cook, A. W., Carter, W. B., and Nidzgorski, F., 1982, Interferon responses of leukocytes in MS, Neurol., 32:104 .

Collins, R. C., Espinoza, L. R., Plank, C. R., Ebers, G. C., Rosenberg, R. A., and Zabriskie, J. B., 1978, A double-blind trial of transfer factor vs placebo in multiple sclerosis patients, Clin.Exp.Immunol., 33:1.

Dachsel, R., Wieczorek, V., and Voigt, W., 1982, Klinische Längschnittuntersuchungen unter immunsuppressiver therapie bei multipler sklerose, Dt.Gesundh.-Wesen, 37:2087.

Dau, P. C., Johnson, K. P., and Spitler, L. E., 1976, The effect of levamisole on cellular immunity in multiple sclerosis, Clin. Exp.Immunol., 26:302.

Dau, P. C., Petajan, J. H., Johnson, K. P., Panitch, H. S., and Bornstein, M. B., 1980, Plasmapheresis in multiple sclerosis: preliminary findings, Neurol., 30:1023.

De Saxce, H., Marteau, R., and Lhermitte, F., 1984, Treatment of progressive multiple sclerosis by the combined application of antilymphocyte serum, azathioprine and prednisone, in: "Immunological and Clinical Aspects of Multiple Sclerosis," R. E. Gonsette, P. Delmotte, eds., MTP Press Limited, Lancaster.

Dominguez, R., Chamoles, N., and Somoza, M., 1978, Intrathecal methotrexate in multiple sclerosis, in: "Humoral Immunity in Neurological Diseases," D. Karcher, A. Lowenthal, eds, Plenum Press, N. Y., London.

Dworkin, R. H., Bates, D., Millar, J. H. D., Paty, D. W., and Shaw, D. A., 1984, Dietary supplementation with polyunsaturated fatty acids in acute remitting multiple sclerosis, in: "Immunological and Clinical Aspects of Multiple Sclerosis," R. E. Gonsette, P. Delmotte, eds., MTP Press Limited, Lancaster.

Ferguson, T. B., Clifford, D. B., Montgomery, E. B., Bruns, K. A., McGregor, P. J., and Trotter, 1983, Thymectomy in Multiple Sclerosis. Two preliminary trials, J.Thorac.Cardiovasc.Surg., 85:88.

Fischer, B. H., Marks, M., and Reich, T., 1983, Hyperbaric-oxygen treatment of multiple sclerosis. A randomized, placebo-controlled, double-blind study, The New England J. of Med., 308:181.

Fog, T., Raun, N. E., Pedersen, L., Kam-Hansen, S., Mellerup, E., Platz, P., Ryder, L. P., Jakobsen, B. K., and Grob, P., 1978, Long-term transfer-factor treatment for multiple sclerosis, The Lancet, I:851.

Fog, T., 1980, Interferon treatment of multiple sclerosis patients. A pilot study, in: "Search for the cause of multiple sclerosis and other chronic diseases of the central nervous system," A. Boese, eds., Verlag Chemie, Weinheim.

Formai, C., Sereni, G., and Zannini, D., 1980, L'ossigenoterapia iperbarica nel trattamento della sclerosi multipla, in: "4th Congresso Nationale di Medicina subacquea ed Iperbarica," Napoli.

Giordano, G. F., Masland, W., Ketchel, S. J., Holland, K, Tilmann, K, Wallace, B. A., and Jones, R. M., 1982, An investigation of lymphocytapheresis in multiple sclerosis, Plasma Ther.Transfus. Technol., 3:417.

Gonsette, R. E., Demonty, L., and Delmotte, P., 1977, Intensive immunosuppression with cyclophosphamide in Multiple Sclerosis, Follow-up of 110 patients for 206 years, J.Neurol., 214:173.

Gonsette, R. E., Delmotte, P., and Demonty, L., 1977, Failure of basic protein therapy for Multiple Sclerosis, J.Neurol., 216:27.

Gonsette, R. E., Demonty, L., Delmotte, P., Decree, J., de Cock, W., Verhaeghen, H., and Symoens, J., 1982, Modulation of immunity in multiple sclerosis: a double-blind levamisole-placebo controlled study in 85 patients, J.Neurol., 228:65.

Gonsette, R. E., Demonty, L., De Smet, Y., and Delmotte, P., 1984, Immunosuppression with cyclophosphamide in multiple sclerosis, in: "Immunological and Clinical Aspects of Multiple Sclerosis," R. E. Gonsette, P. Delmotte, eds., MTP Press Limited, Lancaster.

Guarnieri, B. M., Amaducci, L., Cambi, F., Capparelli, R., Arfaioli, C., Avanzi, G., and Franco, C., 1984, Changes of polymorphonuclear neutral proteinase activity in multiple sclerosis patients before and after plasma exchanges, in: "Immunological and clinical aspects of Multiple Sclerosis," R. E. Gonsette, P. Delmotte, eds., MTP Press Limited, Lancaster.

Guenther, W., Neu, I. S., Koenig, N., and Rothfelder, U., 1984, Experiences with immunoglobulin-G infusions in the treatment of acute multiple sclerosis, in: "Immunological and Clinical Aspects of Multiple Sclerosis," R. E. Gonsette, P. Delmotte, eds., MTP Press Limited, Lancaster.

Hauser, S. L., Dawson, D. M., Lehrich, J. R., Beal, M. F., Kevy, S. V., Propper, R. D., Mills, J. A., and Weiner, H. L., 1983, Intensive immunosuppression in progressive multiple sclerosis. A randomized, three-arm study of high-dose intravenous cyclophosphamide, plasma exchange, and ACTH, The New Engl.J.of Med., 308:173.

Hauser, S. L., Dawson, D. M., and Weiner, H. L., 1983, Immunosuppression for Multiple Sclerosis, letter, New Engl.J.of Med., 309:241.

Hommes, O. R., Lamers, K. J. B., and Reekers, P., 1982, Long-term follow-up after intensive immunosuppression of 47 patients with chronic progressive multiple sclerosis, in: "Immunosuppression in Multiple Sclerosis, Proceedings of the Nijmegen Workshop June 10-12, 1982. Clinical Trials Journal 1984," O. Hommes, J. Mertin, W. W. Tourtellotte, eds, (to be published).

Isomaki, H. A., Mutru, O., and Koota, K., 1975, Death rate and cause of death in patients with rhumatoid arthritis, Scand.J.Rhum., 4:205.

Jacobs, L., O'Malley, J., Freeman, A., Murawski, J., and Ekes, R., 1982, Intrathecal interferon in multiple sclerosis, Arch. Neurol., 39:609.

Johnson, K., 1983, Viruses as initiator of central nervous system autoimmune disorders, in: "Symposium on immune regulation and its application to Multiple Sclerosis, Vancouver, Canada".

Kastrukoff, L. F., McLean, D. R., and McPherson, T. A., 1978, multiple sclerosis treated with antihymocyte globulin - A five year follow-up, J.Canc.Sc.Neurol., 5:175.

Kaudewitz, P., Zander, H., Abb, J., Ziegler-Heitbrock, H. W., and Riethmüller G., 1983, Genetic influence on natural cytotoxicity and interferon production in multiple sclerosis studies in monozygotic discordant twins, Hum.Immunol., 7:51.

Kennes, B., Leroy, C-P., Dumont, J-P., Brohée, D., Jacquy, J., Noël, G., and Nève, P., 1984, Lympho-plasmapheresis in multiple sclerosis: effect on clinical and immunological parameters, in: "Immunological and Clinical Aspects of Multiple Sclerosis," R. E. Gonsette, P. Delmotte, eds., MTP Press Limited, Lancaster.

Kindwall, E. P., 1983, National multiple sclerosis registry, HBO Review, 4:38.

Klaus, G. G. B., 1981, The effects of cyclosporin A on the immune system, Immunology today, 83.

Kolar, O. J., 1983, Clinical application of examination of CSF subpopulation of T cells, in: "International Workshop on multiple sclerosis, Erice."

Lamoureux, G., Cosgrove, J., Duquette, P., Lapierre, Y., Jolicoeur, R., and Vanderland, F., 1981, A clinical and immunological study of the effects of transfer factor on multiple sclerosis patients, Clin.Exp.Immunol., 43:557.

Lauer, K., Firnhaber, W., and John, D., 1984, Longterm treatment with azathioprine in multiple sclerosis patients, in: "Immunological and clinical aspects of Multiple Sclerosis," R. E. Gonsette, P. Delmotte, eds., MTP Press Limited, Lancaster.

Levine, S., and Swoinski, R., 1977, Suppression of the hyperacute form of experimental allergic encephalomyelitis by drugs, Arch.int.Pharmacodyn., 230:309.

McFarland, H. F., and Rose, J. W., 1982, Lymphocytapheresis in the treatment of multiple sclerosis, Plasma Ther Transfus Techn., 3:411.

Massaro, A. R., and Burrai, I., 1981, Levamisole treatment in multiple sclerosis, in: "Excerpta Medica. Proceedings of the 12th World Congress of Neurology, Kyoto".

Mertin, J., Kremer, M., Knight, S. C., Batchelor, J. R., Halliday, A. M., Rudge, P., healey, M. J. R., Compston, A., Thompson, E. J., and Denman, M., 1982, Double-blind controlled trial of immunosuppression in the treatment of multiple sclerosis: final report, The Lancet II:351.

Myers, L. W., Ellison, G. W., Levy, J., Holevoet, M., and Tourtellotte, W. W., 1977, Experience with Levamisole in multiple sclerosis, in: "Excerpta Medica.Proceedings of the 11th World Congress of Neurology, Amsterdam".

Neighbour, P. A., and Bloom, B. R., 1979, Absence of virus-induced lymphocyte suppression and interferon production in multiple sclerosis, Proc.Natl.Acad.Sci.USA, 76:476.

Neighbour, P. A., Miller, A. E., and Bloom, B. R., 1981, Interferon responses of leucocytes in multiple sclerosis, Neurol., 31:561.

Neubauer, R. A., 1980, Exposure of multiple sclerosis patients to hyperbaric oxygen at 1,5-2 ATA: a preliminary report, J.Fla. Med.Assoc., 67:498.

Oger, J. J-F., Antel, J. P., Kuo, H. H., and Arnason, B. G. W., 1982, Influence of azathioprine (Imuran) on in vitro immune function in multiple sclerosis, Ann.Neurol., 11:177.

Pallotta, R., Anceschi, S., and Costagliola, N., 1980, Prospettive di terapia iperbarica nella Sclerosi a Placche, Ann.Med.Navale, 85:57.

Patzold, U., Haller, P., Haas, J., Pocklington, P., and Dreicher, H., 1979, Therapie der Multiplen Sklerose mit Levamisole und Azathioprin, Nervenarzt, 49:285.

Patzold, U., Hecker, H., and Pocklington, P., 1982, Azathioprine in treatment of multiple sclerosis. Final results of a 4½ year controlled study of its effectiveness covering 115 patients, J.Neurol.Sci., 54:377.

Patzold, U., and Haas, J., 1982, Adverse effects of longterm treatment with azathioprine, in: "Immunosuppression in Multiple Sclerosis, Proceedings of the Nijmegen Workshop, June 10-12, 1982. Clinical trials Journal 1984", O. Hommes, J. Mertin, W. W. Tourtellotte, eds., (to be published).

Raine, C. S., and Traugott, U., 1983, Chronic relapsing experimental autoimmune encephalomyelitis. Ultrastructure of the central nervous system of animals treated with combinations of myelin components, Labor. Invest., 48:275.

Romine, J. S., and Salk, J., 1983, A study of myelin basic protein as a therapeutic probe in patients with multiple sclerosis, in: "Multiple Sclerosis," J. F. Hallpike, C. W. Adams, W. W. Tourtellotte, eds., Chapman and Hall, London.

Rosen, J. A., 1979, Prolonged azathioprine treatment of non-remitting multiple sclerosis, J.Neurol.Neurosurg. & Psych., 42:338.

Rosen, A. D., and Hamburger, M. I., 1981, Plasmapheresis in multiple sclerosis: effect on the visual evoked potential, Plasma Ther. Transfus.Techn., 2:239.

Routhier, G., Janossy, G., Epstein, O., Thomas, H. C., Sherlock, S., Kung, P. C., and Goldstein, G., 1980, Effects of cyclosporin A on suppressor and inducer T lymphocytes in primary biliary cirrhosis, The Lancet, II:1223.

Ruutiainen, J., Panelius, M., and Cantell, K., 1983, Toxic effects of interferon administered intrathecally, British Med.J., 286:940.

Sabouraud, O., Madigand, M., and Merienne, M., 1982, Continuous immuno-suppressive therapy for multiple sclerosis: appraisal of 67 cases started before 1972, in: "Immunosuppression in Multiple Sclerosis, Proceedings of the Nijmegen Workshop. June 10-12, 1982. Clinical trials Journal 1984" O. Hommes, J. Mertin, W. W. Tourtellotte, eds., (to be published).

Salonen, R., Ilonen, J., Reunanen, M., and Salmi, A., 1982, Defective production of interferon-alpha associated with HLA-DW2 antigen in stable multiple sclerosis, J.Neurol.Sc., 55:197.

Santoli, D., Hall, W., Kastrukoff, L., Lisak, R. P., Perussia, B., Trinchieri, G., and Koprowski, H., 1981, Cytotoxic activity and interferon production by lymphocytes from patients with multiple sclerosis, J.of Immunol., 126:1274.

Schuller, E., and Govaerts, A., 1983, First results of immunotherapy with immunoglobulin G in multiple sclerosis patients, Eur. Neurol., 22:205.
Seidenfeld, A. M., Smythe, H. A., Ogryzlo, M. A., Urowitz, M. B., and Dotten, D. A., 1976, Acute leukemia in rheumatoid arthritis treated with cytotoxic agents, J.Rheum., 3:295.
Seil, F. J., Westall, F. C., Romine, J. S., and Salk, J., 1983, Serum demyelinating factors in multiple sclerosis, Ann.Neurol., 13:664.
Shih, W. W. H., Baumhefner, R. W., Tourtellotte, W. W., Haskell, C. M., Korn, E. L., and Fahey, J. L., 1983, Difference in effect of single immunosuppressive agents (cyclophosphamide, CCNU, 5-FU) on peripheral blood immune cell parameters and central nervous system immunoglobulin synthesis rate in patients with multiple sclerosis, Clin.Exp.Immunol., 53:122.
Stefoski, D., Schauf, C. L., McLeod, B. C., Haywood, C. P., and Davis, F. A., 1982, Plasmapheresis decreases neuroelectric blocking activity in multiple sclerosis, Neurol., 32:904.
Tourtellotte, W. W., Potvin, A. R., Mendez, M., Baumhefner, R. W., Potvin, J. H., Ma, B.I., and Syndulko, K., 1980, Failure of intravenous and intrathecal cytarabine to modify central nervous system IgG synthesis in multiple sclerosis, Ann.Neurol., 8:402.
Tovell, D. R., McRobbie, I. A., Warren, K. G., and Tyrrell, D. L. J., 1983, Interferon production by lymphocytes from multiple sclerosis and non-MS patients, Neurol., 33:640.
Trotter, J. L., Rodey, G. E., and Gebel, H. M., 1982, Azathioprine decreases suppressor T cells in patients with multiple sclerosis, The New Engl.J. of Med., 306:365.
Turk, J. L., and Parker, D., 1979, The effect of cyclophosphamide on the immune response, J. of Immunopharmacol., 1:127.
Valbonesi, M., Garelli, S., Mosconi, L., Zerbi, D., and Forlani, G., 1981, Plasma exchange in the management of patients with multiple sclerosis: preliminary observations, Vox Sang., 41:68.
Van Haver, H., Lissoir, F., Theys, P., Droissart, Chr., Van Hees, J., Ketelaer, P., Carton, H., Gautama, K., Vandeputte, I., and Vermylen, C., 1984, Transfer factor treatment in MS: a 3 year prospective double-blind study (1982-1985), in: "Immunological and Clinical Aspects of Multiple Sclerosis," R. E. Gonsette, P. Delmotte, eds., MTP Press Limited, Lancaster.
Ververken, D., Carton, H., and Billiau, A., 1979, Intrathecal administration of interferon in MS patients, in: "Humoral Immunity in Neurological Diseases," D. Karcher, A. Lowenthal, eds., Plenum Press, N. Y. and London.
Vervliet, G., Claeys, H., Van Haver, H., Carton, H., Vermylen, C., Meulepas, E., and Billiau, A., 1983, Interferon production and natural killer (NK) activity in leukocyte cultures from multiple sclerosis patients, J.Neurol.Sci., 60:137.
Warren, J., Sacksteder, M. R., and Thuning, C. A., 1978, Oxygen immunosuppression: modification of experimental allergic enc-

ephalomyelitis in rodents, J. of Immunol., 121:315.Weiner, H. L., Dawson, D. M., 1980, Plasmapheresis in multiple sclerosis: preliminary study, Neurol., 30:1029.

Zander, H., Abb, J., Kaudewitz, P., and Riethmüller, G., 1982, Natural killing activity and interferon production in multiple sclerosis, The Lancet, I:280.

Zeeberg, I., Heltberg, A., Kristensen, J. K., Raun, N. E., and Fog, T., 1982, A longterm double-blind, controlled trial - of azathioprine versus placebo, - in treatment of progressive multiple sclerosis, in: "Immunosuppression in multiple sclerosis, Proceedings of the Nigmegen Workshop June 10-12, 1982. Clinical Trials Journal," O. Hommes, J. Mertin, W. W. Tourtellotte, eds., (to be published).

INDEX

Abbreviations:

AM	aseptic meningoencephalitis	LPC	lysophatidyl choline
CNS	central nervous system	MBP	myelin basic protein
CSF	cerebrospinal fluid	MS	multiple sclerosis
EAE	experimental allergic encephalomyelitis	PWM	pokeweed mitogen
		VEP	visual evoked potentials